STEAMstart

Children's Book of
STORIES

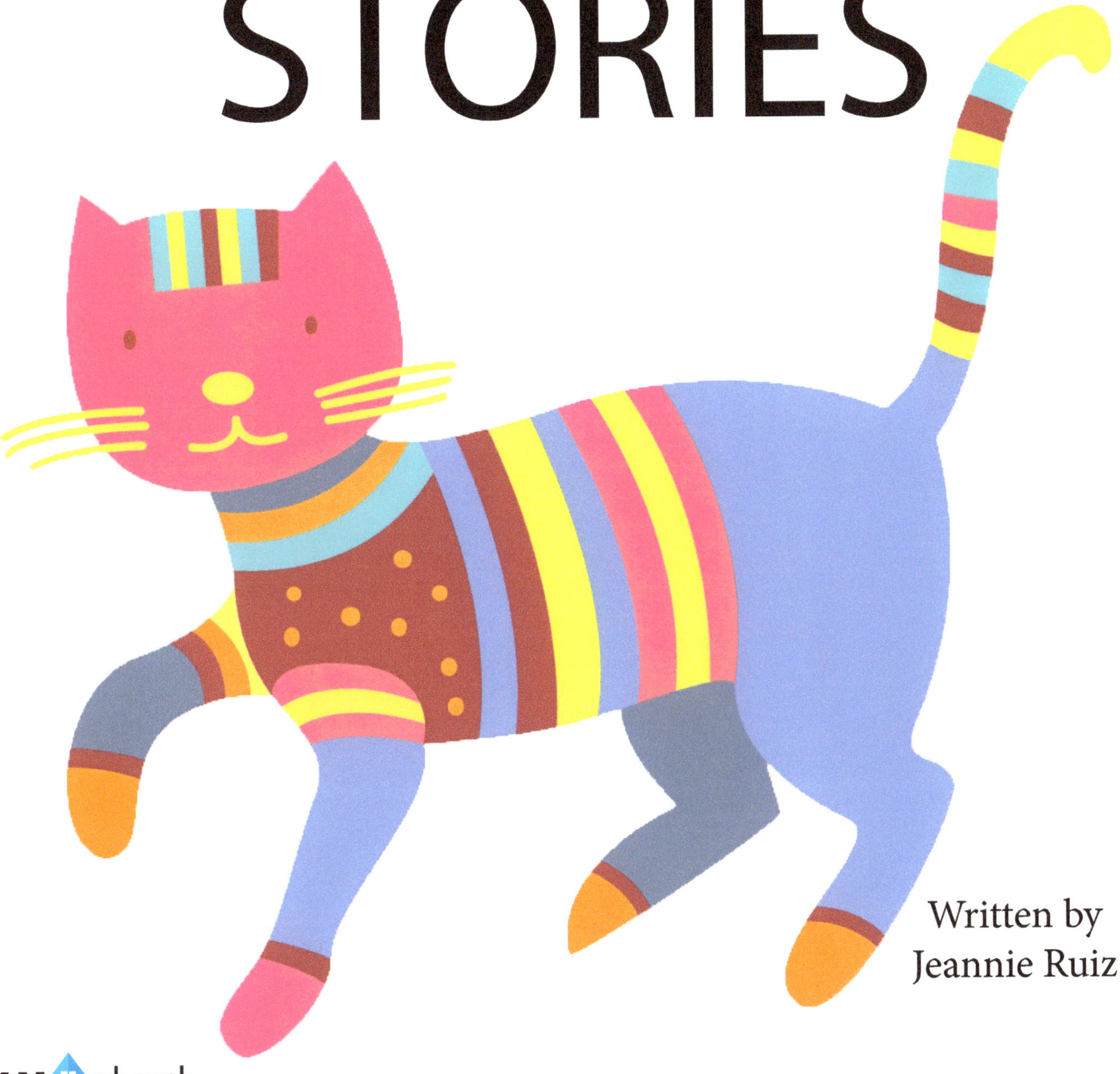

Written by
Jeannie Ruiz

PARTNERS IN STEM PRESS
Portland Oregon

Copyright 2015
by Jeannie Ruiz

Published by PARTNERS IN STEM Press
7401 N Burlington Ave
PORTLAND, OREGON 97203

All rights reserved. No part of this publication may be reproduced, stored in a retrieval system, or transmitted in any form for any reason, recording or otherwise, without the prior written permission of Partners in STEM Press.

ISBN 978-1-942357-31-5

Photo permissions and authors, illustrators, designers, photographers available for download at the URL.

http://www.partnersinstem.org

CERTIFIED READY-TO-SCALE

STEMWORKS EXEMPLARY PROGRAM STATUS

Ten80 Education and its programs are included in the STEMWorks Database by Change the Equation, a non-partisan, CEO-led initiative to connect and align efforts to improve STEM learning. Ten80 Education programs met the highest standards for excellence and exemplary STEM project based programming through rigorous examination by WestEd, an independent nonprofit research organization.

RECOMMENDED BY REAL TEACHERS

We are thoroughly enjoying our STEAM lessons. The program is excellent! My students have loved making the Rhombis and getting to build with shapes and tape. It has been a big hit! Thanks for everything!
Alicia Stenard
Mater Christi in Albany School District

"We need more projects like this that carry on for a longer number of weeks. My teachers used scope and sequence suggestions, and the program ran after school for a semester. We could have worked on this a lot longer."
Karen James
Druid Hills Elementary in Charlotte, NC

"I'm excited for our students to get their hands on real STEAM projects. There isn't much out there for Kindergarten and First Grade."
Sandy Mettler
Fort Worth Schools in Dallas, TX

SECOND GRADE: USING THIS STORYBOOK

JUMP-START INVESTIGATIONS WITH STORIES
Each project and STEAMvestigation has an associated story. The story does not offer the recipe for success in each STEAMvestigation. Rather, the literacy component is a jump-start to discussion on the topic. Images are limited to encourage kids to offer their own vision. The goal of this book is to encourage the "bedtime story" effect. They will interact with plenty of books that walk them through with pictures. Let students tell you what they think Quad looks like or what a Saturday Farmer's Market might sell.

THESE STORIES ARE WRITTEN ABOVE GRADE LEVEL
These stories are intentionally written above the level of a first grader's reading ability. Children will not build mature syntax or extensive vocabularies if they are limited to six-word sentences. Building the ability to "see" a story can only happen when the child hears the story and begins to imagine.

READ AND RE-READ
Kids don't get bored with repetition in television or books. Read the same book several days in a row. Hold a discussion each time about the images, or let kids draw their own versions. Use this book just like you'd use any compilation. You are the teacher. You know when the storybook read-aloud component is appropriate for your students.

S.T.E.A.M. IN THE CONTEXT OF SCIENCE CLASSROOMS
Recurring themes are pervasive in sciences, mathematics, and technology. These ideas transcend disciplinary boundaries and include patterns, cycles, systems, models, and change and constancy.

KINDERGARTEN
- Students observe and describe the natural world using their five senses.
- Doing science as inquiry develops and enriches understanding of scientific concepts and processes.
- Students develop vocabulary while investigating common objects, earth materials, and organisms.
- Counting, place value, numeracy, and measurement are given useful context within the scope challenges.
- Students should begin to recognize math as the language of science.

FIRST GRADE
- Students observe and describe the natural world using their five senses.
- Students develop/enrich abilities to understand the world in the context of scientific concepts and processes.
- Students develop vocabulary investigating properties of common objects, earth materials, and organisms.
- Students are encouraged to use their budding mathematical skills as useful tools in the pursuit of solutions.
- As designers, students will activate these skills in inquiry-focused projects.

SECOND GRADE
- Students observe and investigate to learn about the natural world and reveal patterns, changes, and cycles.
- Students should understand that certain types of questions can be answered by using observation and investigations and that the information gathered in these may change as new observations are made.
- As students participate in inquiry-driven and project-based investigation, they begin to understand the integration of mathematics in the context of scientific investigation and the design process.

SECOND GRADE MODULE 1: CONTENTS

ROUND AND ROUND		6
SCIENCE	The Adventures of Luna and Sol	8
TECHNOLOGY	Rover, Rover, Come on Over	10
ENGINEERING	Monster Wheels	12
ARTS	Quilting in Circles	14
MATHEMATICS	Radius and Pi	16
LOOKING FOR SQUARES		18
SCIENCE	Square Foot Harvest	20
TECHNOLOGY	P Is for Parallel	22
ENGINEERING	Shopping for Shapes	24
ARTS	Stories in the Square	26
MATHEMATICS	Paper Forests	28
TASTY TRIANGLES		30
SCIENCE	Rickety Rackety	32
TECHNOLOGY	City of Three Angles	34
ENGINEERING	Traveling Triangle	36
ARTS	Fox's Forest	38
MATHEMATICS	Tri, Tri, Again	40
FLYING FIVES		42
SCIENCE	We're Hunting for a Pentagon	44
TECHNOLOGY	Tool Time	46
ENGINEERING	Creative Design	48
ARTS	Design Time	50
MATHEMATICS	The Pentagon, USA	52
ON THE MOVE		54

2D SHAPES

 1
 2
 3
 4

EXTENSION

ROUND AND ROUND
Explore circles.

Once upon a time, there was Rhombi. Rhombi loved shapes and found them everywhere. She especially loved the circles she saw in the moon and sun.

On many nights, Rhombi sat in a meadow to watch the moon. She noticed that its shape appeared to change from week to week. Rhombi wanted to see if the moon's shape really changed or if there might be some other explanation.

Rhombi's uncle had given her a telescope for her birthday. Telescopes used lenses shaped like circles to make faraway objects look larger and closer.

Rhombi and Uncle Reel watched a show about the movement of planets and the revolution of the moon around the Earth. Rhombi saw giant ellipses and circles moving through the Solar System.

Uncle Reel and Rhombi headed to the rooftop telescope. He adjusted the round lenses to help Rhombi see clearly. The moon came into focus. Light from the Sun shone on its face, helping Rhombi see it full and round. It was the coolest circle she'd ever seen.

Rhombi decided that she wanted to share her view of the sky with friends. The astronomy club was having a festival, and she could set up a place to view the moon. How would she move the heavy object from place to place? Rhombi looked out beyond the pages of her world and asked, "Designers, would you help me create and build a cart to move my telescope?"

The Adventures of Luna and Sol
Explore circles.

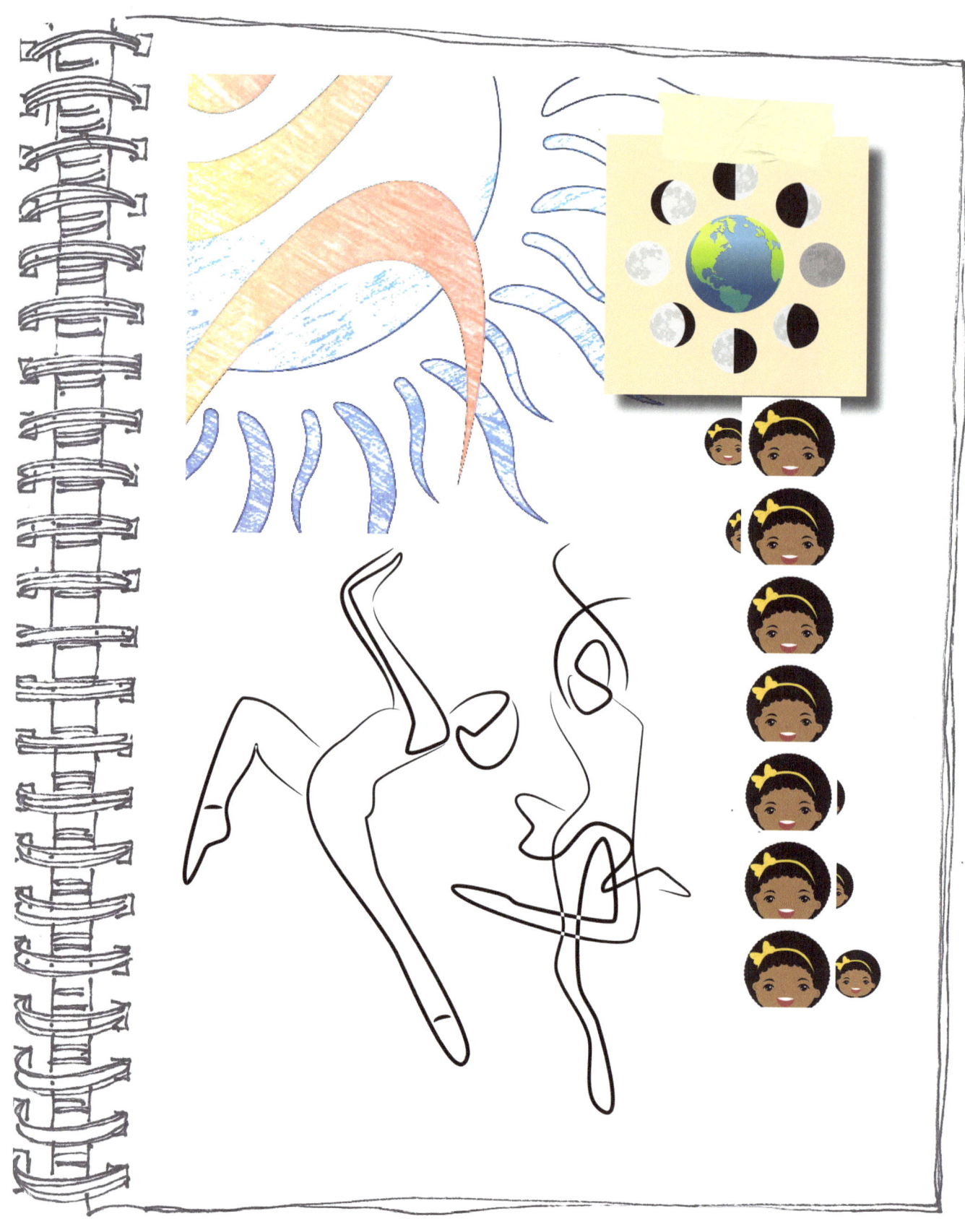

The Adventures of Luna and Sol
Explore circles.

Every day was an adventure for Luna. She loved to travel and never, ever sat still. Her favorite dance in the world was a twirling rotation. Like a ballerina on a jewelry box, Luna could spin and spin and spin.

Even though they were different, Luna and Sol stayed in touch no matter how much Sol traveled. Sol kept an eye on Luna, warming her with his fiery light. Luna had no light of her own. The side of Luna that faced away from Sol was always cold and dark. But the side that faced him was warmed by Sol's light. Each month - well, closer to 29 and a half days - Luna tuned around one full turn.

Luna's whole life was a dance. Big Sister Earth danced around Sol while Luna danced ahead and sometimes behind her, always smiling when Sol poured his light and warmth on her face.

Each month, as they moved through the First Quarter of the dance, Luna twirled between Big Sister Earth and Sol,

Big Sister Earth could not see Luna. When that happened, Luna and Sol smiled at one another and Luna twirled faster so Big Sister Earth could start to see her again.

Sol was so big that he cast light on Luna, even if Big Sister Earth was in the way.

Sometimes, Big Sister Earth managed to dance between them so that Luna was in her shadow and Sol could not see Luna.

That did not happen very often.

Luna would giggle and peek out at Sol after a few minutes. Little by little she moved back into Sol's view and continued to dance.

Rover, Rover, Come on Over
Explore circles.

Two rovers sit still on the moon's surface. The dust on their roofs had not moved in many years. No wind had stirred the little flags attached to the rover antennae. There was no atmosphere on the moon to create weather or wind.

That's Heavy, Man
The LRVs or Lunar Rover Vehicles were very delicate and could never be driven on Earth. They were specially designed for the moon's gravity. Earth's gravity is much higher, making everything on Earth feel heavier than those same objects would feel on the moon.

Apollo Missions 15, 16, 17
The two rovers had been used many times in three lunar visits. Apollo 15, Apollo 16 and Apollo 17 astronauts used the same rovers, parking the vehicles between missions. The Lunar Rover Modules never drove farther than 4.72 miles from the rocket's landing place.

4.72 miles

Hammer Time
On the Apollo 17 mission, the hammer in Commander Gene Cernan's space suit pocket accidentally broke one rover's fender. If astronauts hadn't repaired the fender, moon dust would have clogged delicate machinery and equipment

Rover, Rover, Come on Over
Explore circles.

Duct Tape Designs
On the moon, there are no mechanics to fix a fender, so the astronauts had to use materials they had on hand. The Commander and his team created a new fender using a laminated map and duct tape. You could see the actual EVA* map fender on display at the National Air and Space Museum.

Men on the Moon
The rovers have sat still and silent for more than 40 years. The last visit to the moon launched on December 7, 1972. At the end of that mission, the rovers were present as astronauts placed a permanent plaque on the moon.

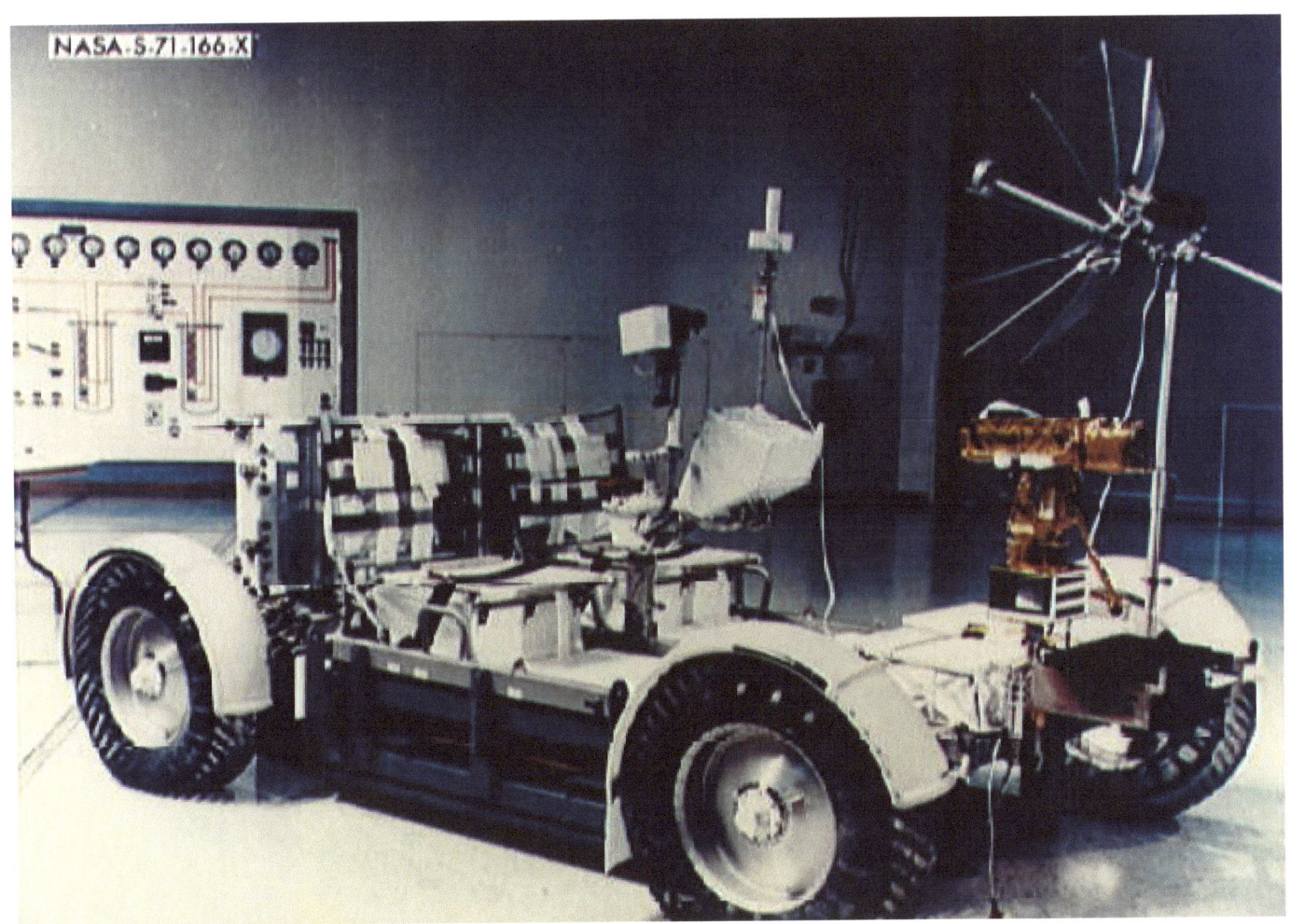

"A17-plaque" by NASA - http://images.jsc.nasa.gov/lores/S72-55417.jpghttp://science.ksc.nasa.gov/mirrors/images/images/pao/AS17/10075902.jpg. Licensed under Public domain via Wikimedia Commons - https://commons.wikimedia.org/wiki/File:A17-plaque.JPG#mediaviewer/File:A17-plaque.JPG PUBLIC DOMAIN

According to Space.com, the rovers measured about 10 feet long (3 meters), 6 feet wide (nearly 2 meters) and almost 4 feet (1 meter) high.

Monster Cart
Explore circles.

Monster Cart
Explore circles.

Radius and Pi were best friends, but they were different people. Radius loved moving fast, talking quickly, and diving into a project. Pi was a planner. Pi made lists, thought about her words before speaking, and moved gracefully.

Radius and Pi did everything together, and they made a good team. Every Saturday, the two pulled their cart along the path to sell vegetables and fruits from their garden at the local Farmer's Market. People set up tables under little tents to sell what they'd grown or made at home. Radius and Pi were popular. They never missed a Saturday market.

This Saturday, though, Radius and Pi had a problem. Rain had started Wednesday and stopped just after dark on Friday night. Water had mixed with dirt on the paths, and the cart's thin tires were sinking more than 3 inches into the mud.

Pi sat down to think about the problem. She pulled out a notebook and started a list. Pi thought of all the vehicles she knew that could move through slippery muck. So far, her list had only two items: monster trucks and sleds.

While Radius dove right into his project, Pi peered into the tool shed looking for ideas. She spied a couple of old pedal toys with big, wide wheels. Pi knew that monster trucks had wide tires with deep grooves in the rubber. The pedal toy's tires were at least 4 inches wider than the cart tires.

Radius helped Pi prop up the cart. They used a screwdriver and wrench to remove the cart's wheels. Radius held the cart steady as Pi connected each wheel to the cart's axle. The twins lowered the cart to the ground. Radius gently pulled the handle. He tugged a bit harder. Finally, the cart started moving. The new, wider wheels worked pretty well! Their cart sank only about 1 inch into the mud. Pi and Radius headed to the market with the Monster Cart.

Quilting in Circles
Explore circles.

Quilting in Circles
Explore circles.

Radius and Pi were visiting Aunt Bee. She had promised to teach the twins how to quilt, and they were both excited. Aunt Bee's quilts were works of art.

Aunt Bee had already made the basic part of the quilt. It had three layers: the bottom layer was sewn in squares of old t-shirt and clothes fabric; the middle layer was fluffy cotton batting, and the top layer was bright fabric that would show the final pattern. Aunt Bee set up a quilt hoop with one hoop under the fabric and another hoop over the fabric. She tightened the hoops around the fabric with a small nut and screw. It drew the fabric tight so needles could move through from top to bottom without bunching the layers. Bee gave each twin a thimble to protect their fingers from sharp needles as they worked.

Radius and Pi had already used a rotary cutter with a sharp round blade and a template to cut circles in several sizes. They had measured the circles from each shape's center to its outer edge. Radius chose to cut circles that measured 1 inch from center, 2 inches from center and 3 inches from center. They had 22 circles in different colors to sew.

Pi had drawn a pattern of circles that made lots of dots on the fabric. Some of the circles touched edge to edge. Other circles sat one inside another in a quilting pattern called an echo. They hand-sewed their circles one by one, moving the hoop each time they finished an area of the quilt.

The planet was rotating away from the Sun when Radius and Pi finished their project. Just as the Sun vanished over the horizon and the moon was visible in the sky, the twins put down their needles and shook out the beautiful quilt. Circles overlapped and danced on one side while brightly colored and patterned squares filled the back. The quilt would always remind them of a great day with Aunt Bee.

Radius and Pi
Explore circles.

Radius and Pi
Explore circles.

Radius and Pi rolled along the path toward Saturday market. Suddenly, the cart stopped. Radius circled the cart looking for a problem. He leaned down to take a look at the wheel and axle. The wheel had wobbled loose and cracked.

Radius looked around for some way to fix the cart. The two needed a new wheel. Radius and Pi began looking for round objects that could fit the cart. The two found a roller skate wheel and a skateboard wheel in a pile by the curb. Both wheels would be too small. The length from the center hole to the outer part was just too short. Their cart might roll, but it would tilt and spill the apples.

They found some balls, but none had a hole in the center. No hole meant no axle. How would the ball stay on without an axle down the middle?

Radius tried to hold up the no-wheel corner while Pi steered. The two entrepreneurs really wanted to reach the market in time to sell their fruit. They moved too slowly without a wheel.

Pi and Radius looked at the bicycles in a shop window. They looked at each other with wide grins and carefully propped the cart on a step. Four minutes later, Radius emerged with a bicycle wheel minus the tube. It was old and rusted but not bent. The shop owner gave them the wheel. The wheel was just the right length from center to rim.

Radius and Pi put the wheel on their cart. They zoomed to the market. When the market gate opened, customers were happy to see the two. Radius and Pi sold all the yummy ripe apples. The two were proud that they'd solved an engineering problem all on their own.

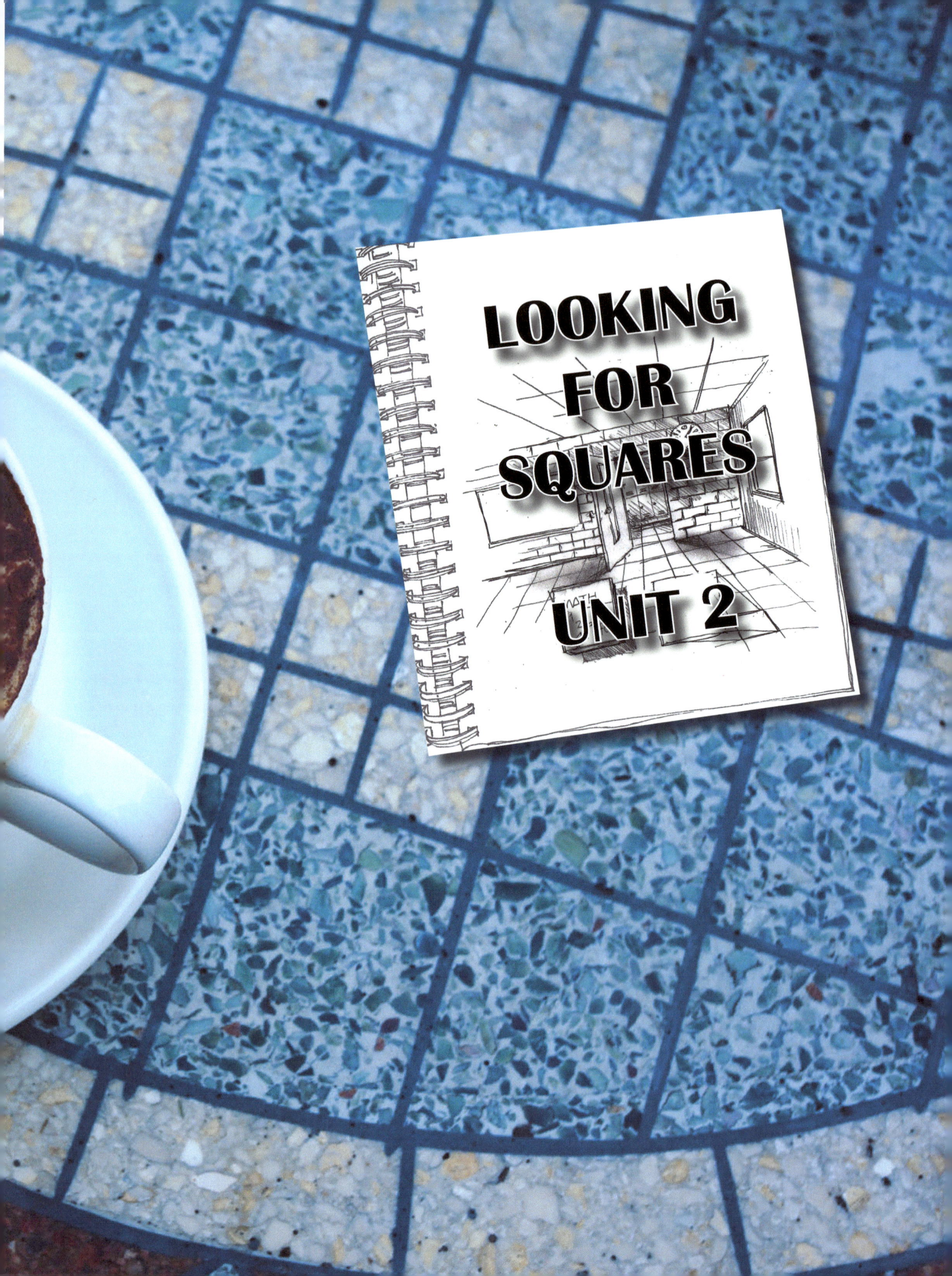

LOOKING FOR SQUARES
Explore squares.

Once upon a time, there was Rhombi. Rhombi loved to travel, and she had recently visited a planetarium to see the moon. She had carried the telescope from that visit to share at festivals.

Rhombi had loved the circular shape of the moon when it was full. She wondered what other shapes were hiding in plain sight.

Rhombi decided to take a walk around the school to find more shapes. Inside her school, Rhombi found a round clock on the wall. She saw a picture of girls playing hopscotch on square boxes.

Rhombi found many square-looking shapes in her classroom. The floor tiles looked like squares. The ceiling tiles looked like squares. Even the math book looked like a square.

Rhombi knew that not all square-looking things were really squares. Rhombi took a closer look. She knew that a square had four equal sides with four corners shaped like the letter L.

The tabletop in her classroom needed some new tiles. The tiles needed to be true squares. How would Rhombi decide which tiles at the store were real squares?

Rhombi needed a tool to prove that things were squares or some other quadrilateral. Just staring at the shapes wasn't proof, and Rhombi didn't want to waste money buying tiles that wouldn't work on her tabletop.

Rhombi looked out from the pages of her world and asked, "Designers, would you help me create a tool to help identify squares?"

Square Foot Harvest
Explore garden harvests in science.

Square Foot Harvest
Explore garden harvests in science.

Leaves crunched underfoot as Rhombi headed toward the garden. She kicked a dozen colors into the air just to watch them float back to the ground. Fall was her favorite part of the year.

Squirrels chittered as they collected acorns from the ground under oak trees more than 70 feet tall. The trees towered above Rhombi like a giant swaying ceiling. She knew these trees must be at least 50 years old if they were dropping so many acorns.

Rhombi and her friends had planted vegetable seeds in the Community Garden in August. Over the past 2 months, the seeds had germinated, pushing out of their hard seed covers. Seedlings began to push through the soil toward the sunlight. Rhombi had spent many hours thinning the rows so each seedling would have enough sun and room to grow. When she saw leaves, Rhombi knew that the radish bulb and roots were reaching down into the soil below. Finally, after almost 40 days, the radishes were ready to eat.

Carrots and radishes and lettuce waited for Rhombi. She pulled 20 carrots and put them into her basket. Next she layered 24 radishes. Finally, Rhombi added 4 heads of lettuce. She picked up her heavy basket and headed home through the crunchy leaves.

She washed and chopped carrots and radishes. She tore lettuce leaves, and added them to a big bowl. When her friends sat down to dinner, Rhombi would proudly serve her delicious autumn harvest salad!

P is for Parallel
Explore shapes in technology.

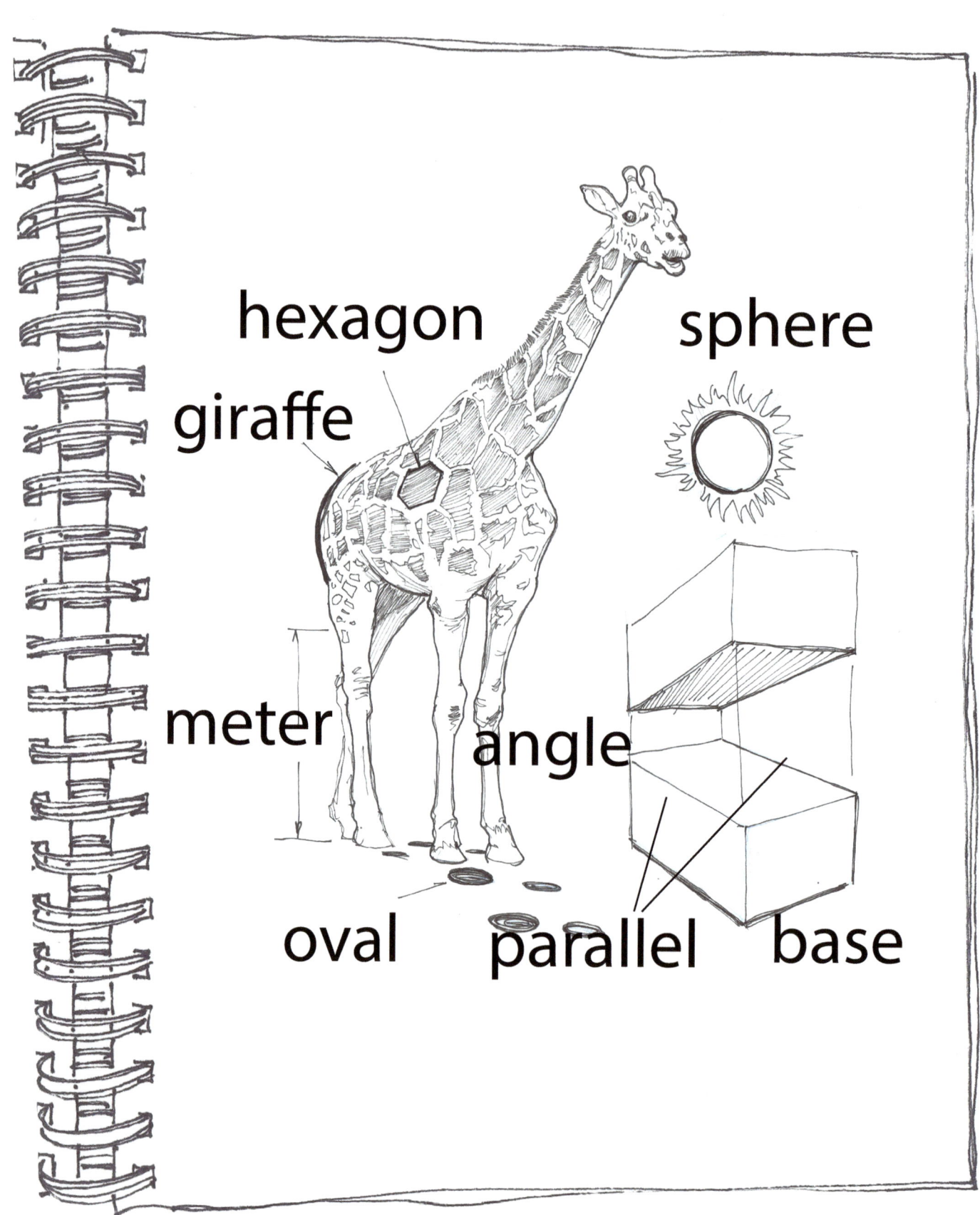

P is for Parallel
Explore shapes in technology.

A is for Angle, where two lines meet up...
B is the Base, a 3D shape's bottom or top.
C is for Coordinate, where x and y are.
D is for Decagon, like the drawing of a star.
E is for Ellipse that looks like an oval.
F is for Fractal, every part like the whole.
G is for Google, 1 with 100 zero's.
H is for Hexagon, the honeycomb's hero.
I is for Interior including all the stuff inside.
J is for Junction where things touch or collide.
K is a Kite with four triangles glued together.
L is for Line. Does it goes on forever?
M is for Meter, a unit of measure.
N is for Nonagon, a nine sided treasure.
O is for Octagon. Octa means 8 sides.
P is for Parallel, two lines that never collide.
Q is a Quadrilateral, 4 sides and 4 corners.
R is for Rotation, to turn around a shape's center.
S is for Spheres that are perfectly round.
T is the Triangle that keeps a bridge from falling down.
U is for Undecagon, a 10-sided shape with no order.
V is for Vertex, hanging out at the corner.
W is for Width, a measure from side to side.
X is for X-Axis, a horizontal line.
Y is for Y-Axis, up and down on a graph.
Z is for Zero, no dinner for giraffe?

Shopping for Shapes
Explore shapes in engineering.

shopping list

cheese   fruit
cookies gum
crackers

Shopping for Shapes
Explore shapes in engineering.

Poly and her friend Trap were shopping for groceries. The two were meeting up with Rhombi that afternoon for chess club. It was Poly's turn to bring snacks. She had ten dollars to spend.

Poly pushed a cart up and down the aisles looking for something her friends might like to eat. She dropped a pack of gum in the cart, but it fell through the squares that covered the cart. She placed the gum on the rectangular face of a box of cookies.

Poly wanted to find cheese, crackers, fruit, and cookies. So far, she'd found cookies. They were the round kind with chocolate chips. Trap held up a bag of $3.00 star fruits. They looked like spheres or balls now. When cut, the star fruits would look like little green star shapes on a plate. He also had fat purple grapes that cost $1.00 per bag.

The list included cheese, crackers, fruit and cookies. So far, Poly and Trap had cookies and fruit. The two friends headed toward the dairy section to look for cheese. Wow, there were a lot of cheese shapes. Which one should they choose? There were cylinders of string cheese. Poly pointed to brick shaped rectangular prisms of cheddar. Trap spied a bag of already cut cubes that would be easy to pile on a tray. The $2.00 cheese cubes were added to the cart.

The last item on their snack list was crackers. The cracker aisle was filled with boxes of all shapes and sizes. Oval buttery crackers sat next to 2 inch square saltines and little ½ inch round soup crackers. There were whole wheat rectangles woven like blankets and cheesy 1 inch squares. Some were even shaped like goldfish. Trap picked the $2.50 woven rectangles. They tasted good with cheese.

Their cart was filled with delicious shapes. Star fruits, grapes like spheres, cubes of cheese, and square crackers filled the cart. The total cost was $8.50. They had one dollar and 50 cents left. Poly and Trap took their bags and headed to the chess club. Snack time shaped up to be a yummy break.

Stories in the Square
Explore squares and patterns in art.

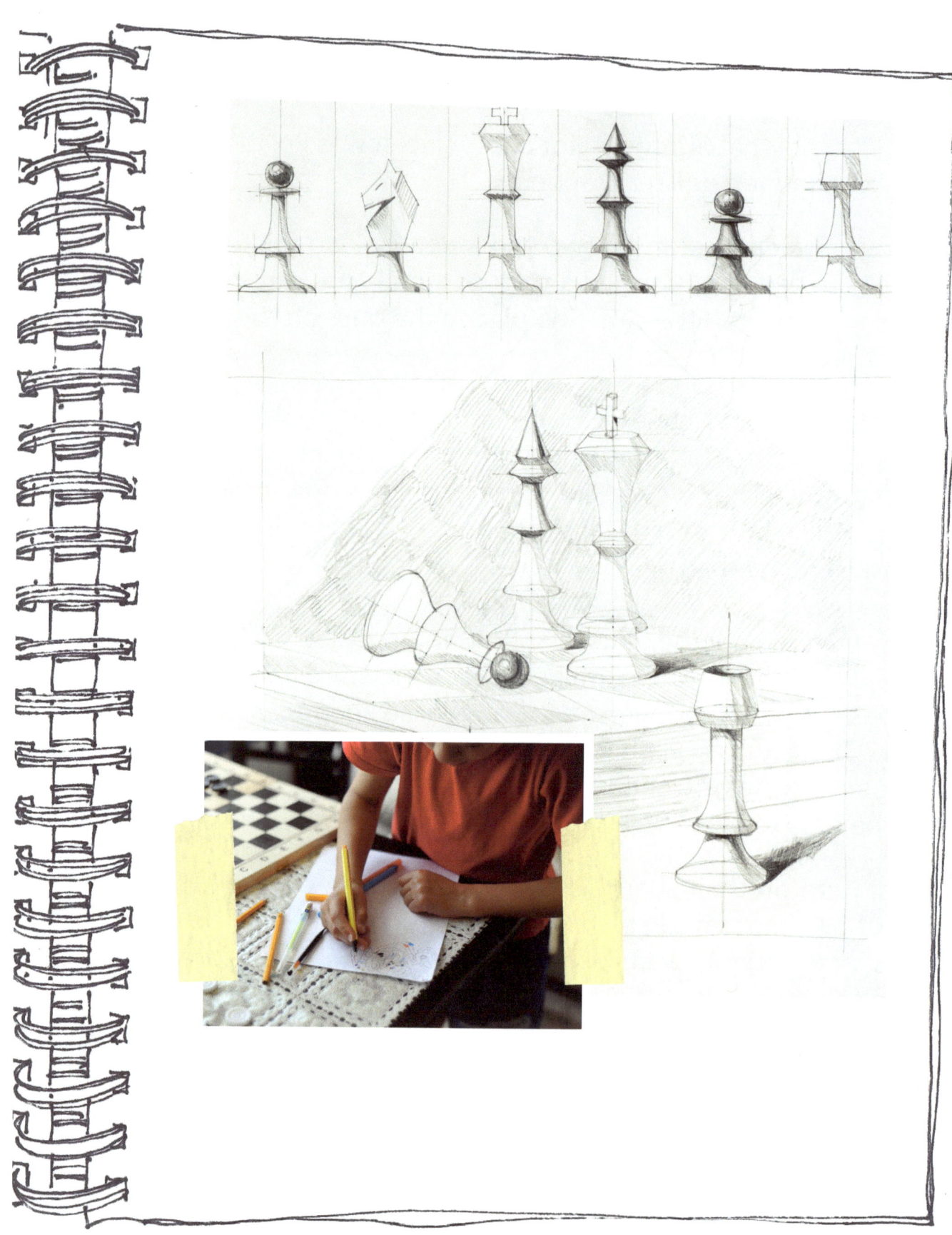

Stories in the Square
Explore squares and patterns in art

Quad had a problem. His class was making a quilt for the autumn fundraiser. Each student was responsible for creating a pattern and making one square called a block. All the blocks would be sewn together to make one big quilt. He had no idea what to make.

His floor was littered with crumpled papers. Quad could not choose a pattern. He had drawn and colored and traced until his fingers hurt, but none of the drawings seemed right for his piece of the class quilt. What was Quad going to do? His brain needed a jump-start.

Quad put down his pencil and headed out to chess practice. He would think more about the quilt block later. At the recreation center, Quad sat down at his usual table. Rhombi sat across the chess board, placing pawns in a row. Behind the 8 pawns, she placed 1 king, 1 queen, 2 rooks, 2 bishops, and 2 knights on the board. The kids took turns moving pieces across the 64 black and white squares. Rhombi almost always played red, while Quad played the white pieces.

As they played, Quad got an idea. He hurried through the match, said a quick goodbye to Rhombi, and hurried home. Grabbing a new piece of paper, Quad folded a square paper in half 3 times. Opening the paper, Quad counted 8 columns. He folded the paper in half 3 times in the other direction. Now he had 64 equal squares.

Quad colored every other square black. He added 8 red pawns, 1 red king, 1 red queen, 2 red rooks, 2 red bishops, and 2 red knights to the black and white board. Quad used his pattern to cut the fabric for his quilt block. He carefully sewed all the pieces together and held the finished quilt block to take a look. It was a really cool pattern.

When the class quilt was pieced together, Quad proudly added his chess board. That week, at chess practice, Quad asked Rhombi for a favor. She agreed with a laugh as Quad set up the Red chess pieces and got ready to play.

Paper Forests
Explore origami in mathematics.

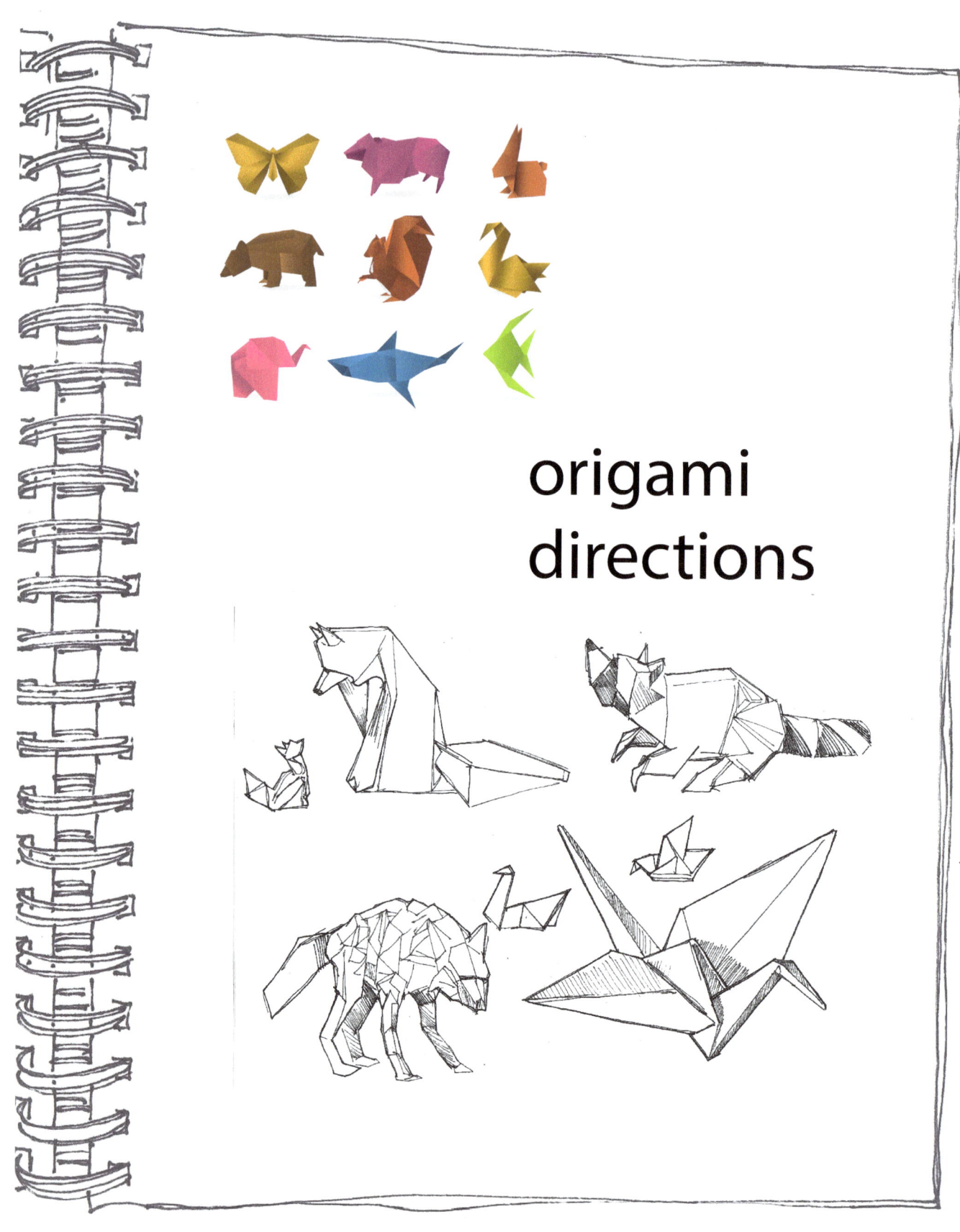

origami directions

Paper Forests
Explore origami in mathematics.

Rhombi's Grandpa folded beautiful paper art. When Rhombi was little, Grandpa could make animals and flowers and boats and castles appear from his hands using nothing but squares of thin paper. Grandpa taught her to fold paper boxes to hold paper jewels. He taught Rhombi to make beautiful princesses with swords that rescued dragons from wicked wizards. Rhombi's favorite project was a mobile hung with 40 folded animals. Grandpa called it his forest in the sky.

Grandpa no longer folded paper, though. His hands were old and gnarled. The fingers couldn't hold delicate paper anymore. Now, instead of the country, Grandpa was living in a city home with lots of other grandmas and grandpas. They all played checkers and had holiday parties. Grandpa seemed happy, but sometimes he talked to Rhombi about the old days. They laughed about a funny fast squirrel that sneaked in through the window and raced around the bedroom before bursting out the front door.

Grandpa remembered pushing Rhombi on the tire swing. He talked about animals that visited from nearby woods. Grandpa especially missed the red fox that used to peer in through the bedroom windows on cold snowy nights. Sometimes they found it sleeping in his shed. Grandpa's new home was in the city. No animals peeked in city windows.

Rhombi planned to visit Grandpa for his birthday on Saturday. Sitting at her desk, Rhombi thought about a gift. She knew he missed his forest and fields, but Rhombi could not bring any live animals into Grandpa's new home. Looking at all the origami papers on her table, Rhombi decided that she could bring some animals to Grandpa after all. She picked up a stack of papers and started to fold.

Saturday was the big day. Rhombi eased her gift through the front door of Grandpa's building. Other grandmas and grandpas and aunties and uncles smiled as she made her way to the dining room. All Grandpa's friends were there with cake and balloons. Everyone oohed and ahhed as Rhombi entered. Grandpa turned to look, and his hands flew to his mouth. He smiled. He grinned. He laughed out loud!

Hanging from Rhombi's hands was a mobile full of forest animals. His squirrel was there in its tree. There were blue birds and rabbits. But, best of all, hanging bright and sassy at the top of Grandpa's new forest in the sky was a beautiful bright red fox. Happy birthday, Grandpa!

TASTY TRIANGLES
Explore triangles.

Once upon a time there was Rhombi. Rhombi loved shapes. Circles were cool (there were circles in the word "cool"). She had also had help figuring out how to test for squares, and that was a lot of fun.

Rhombi's pen-pal from Japan had sent square origami paper and instructions for making a folded fan. She had just accordion folded the longest side of the paper when saw a tiny commotion down by the stream.

3 tiny mice were trying to haul a tiny box of raisins across the park to their mouse hole. They were stuck with no way to cross the stream. Rhombi wished she could help.

When Rhombi returned to her park bench, someone had dropped the Wednesday newspaper on top of her fan. Oh no! Rhombi expected the fan to be smashed flat, but it was not smashed at all.

The folded paper was holding up the newspaper all by itself.

Rhombi bent down to look more closely at the paper. Once folded, the paper fan had made strong triangles with the bench as each triangle's base.

Rhombi looked around the park. She saw triangles in the bridge. She saw triangles under the park bench. Rhombi even saw triangles holding up the bicycles ridden in the park.

Rhombi thought about the little mice and their task. She wished they had a bridge. Wait! Rhombi new how to help the mice. She looked down at all the origami paper in her hands. The mice needed a bridge.

Rhombi would need help, and she knew just where to find the helpers. She looked out beyond the pages of her world and asked, "Designers, can you create a folded paper bridge to help the mice get their raisin box home?"

Rickety Rackety
Explore structures in mathematics.

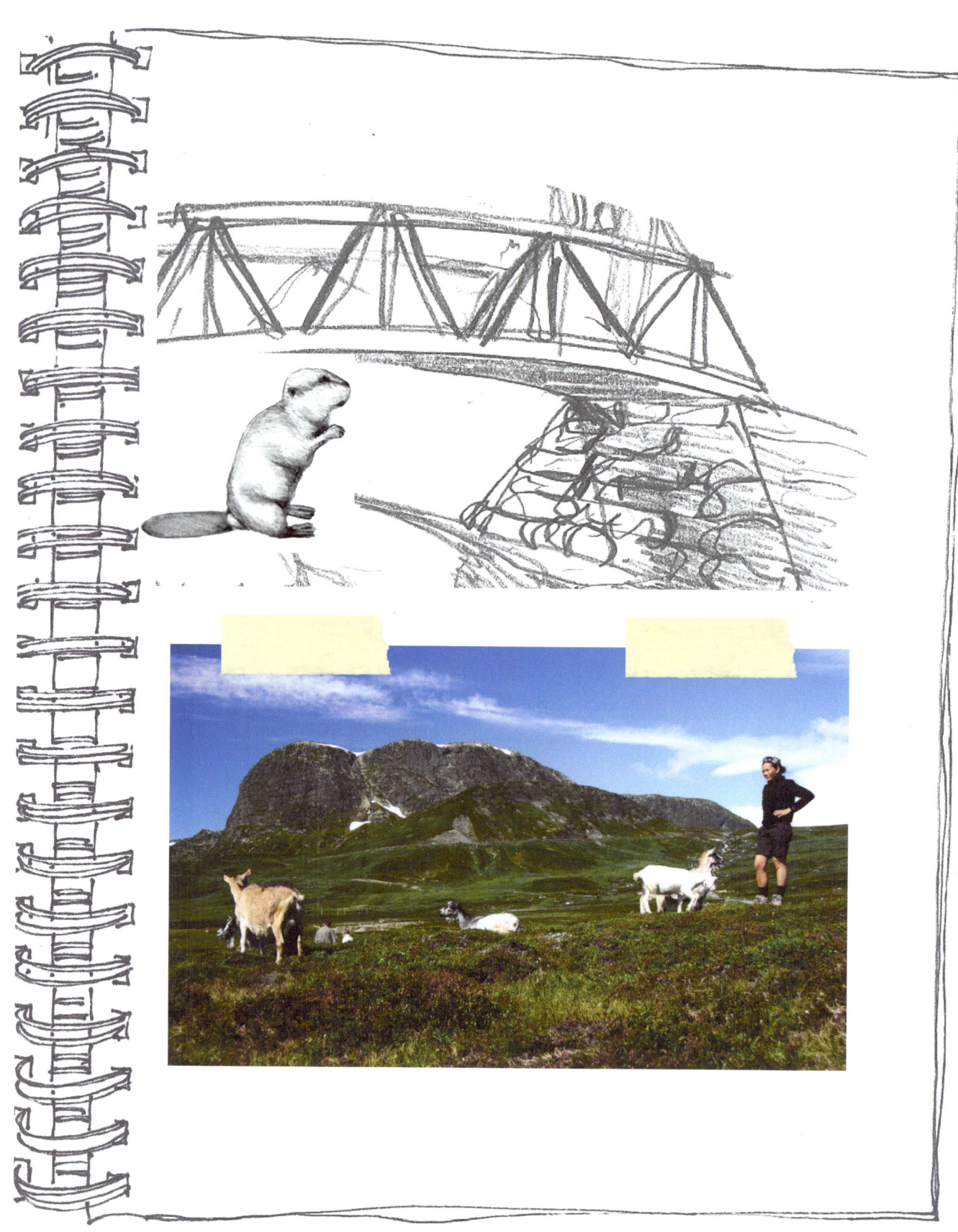

Rickety Rackety
Explore structures in mathematics.

Once upon a time there were three little goats. They were hungry and wanted to eat clover at the top of the triangle hill. To reach the clover, each goat had to cross a rickety rackety bridge. Youngest goat began to cross the rectangle bridge. He dashed across the 12 feet of bridge as it swayed back and forth under his weight. Rickety rackety rickety rackety went the bridge. The beaver under the bridge shouted at them to go away. "You are too loud, and you're making my bridge go rickety rackety."

Youngest Goat answered, "I hope the square-bottom bridge will hold. My brother is on the way, and he is much bigger than I."

After 30 minutes, Middle Goat began to cross the bridge. Rickety rackety rickety rackety went the bridge. The beaver under the bridge lumbered out and shouted at them to go away. "You are even louder than your brother, and you're making my bridge go rickety rackety."

Middle Goat answered, "I hope the square-bottom bridge will hold. My brother is on the way, and he is much bigger than I."

Along the path came Biggest Goat. He took one look at the bridge and shook his head slowly. His triangular coal black beard waggled back and forth. The biggest goat put one hoof on the bridge. Rickety rackety rickety rackety went the bridge. The angry beaver underneath shouted at them to go away! The goat peered under the bridge and saw that the bridge was held up by a single square. He looked across the stream at his brothers. If he couldn't fix the bridge, the younger kids would be stuck on the hillside. Biggest Goat thought and thought. What was the strongest shape he could remember?

Youngest Goat saw a seed and bleated, "What about a circle?"

Biggest Goat shook his head. The bridge would roll away. Middle Goat saw a flower petal and cried, "What about an oval?"

Biggest Goat shook his head. The bridge would rock back and forth like a rocking horse. He needed a very powerful shape to keep the bridge straight and strong. Finally, Biggest Goat looked up at the mountain that held the delicious clover. What shape is a mountain? It's a triangle. Worn by weather and wind, the mountain stands strong for thousands of years. Biggest Goat asked the beaver to help him with the bridge. The beaver grumbled but made lots of triangles to hold up the bridge. Finally, Biggest Goat began to cross the bridge. The happy beaver waddled back under his very strong bridge. He didn't even notice when the three brothers headed back across the bridge at sunset.

City of Three Angles
Explore maps in technology.

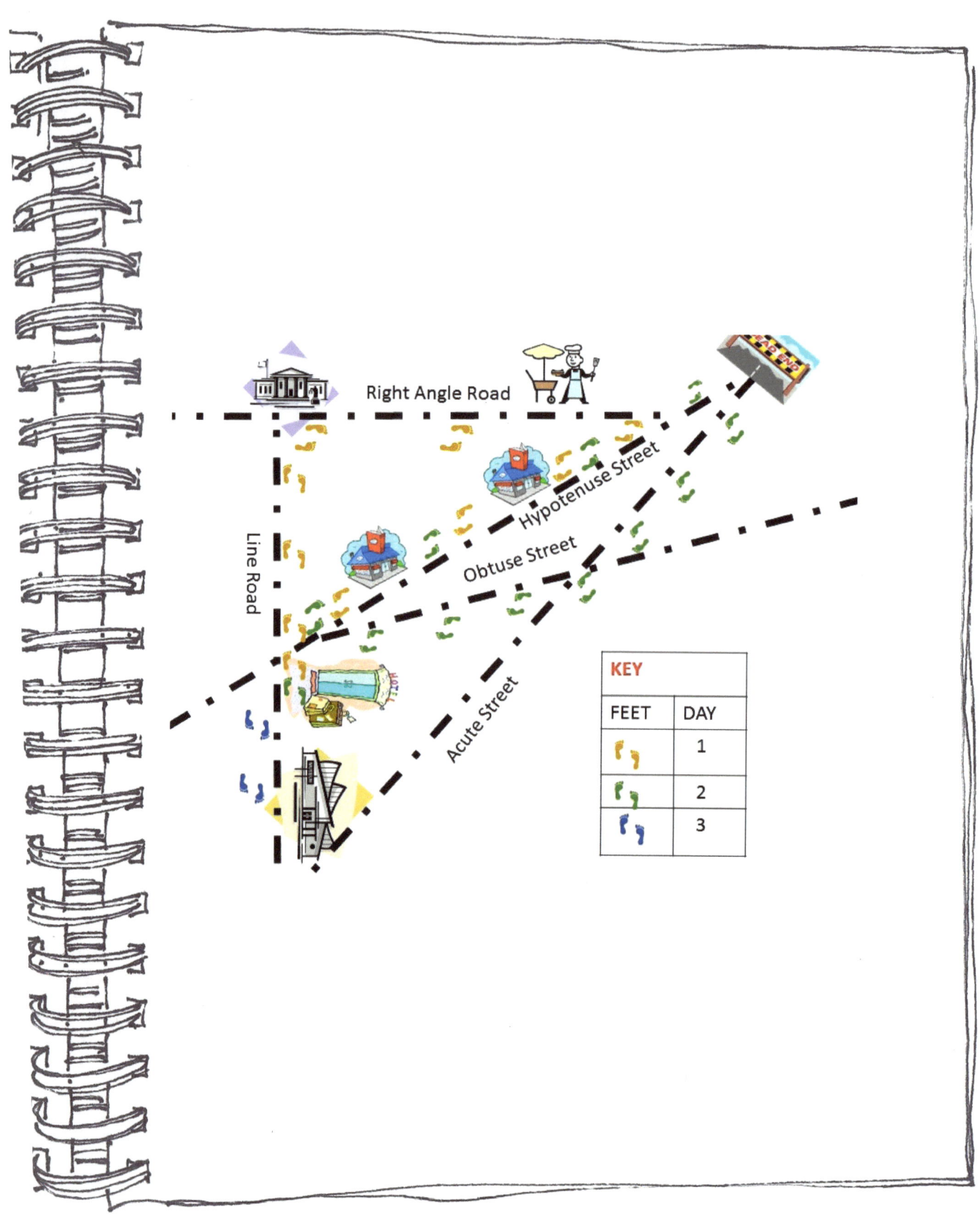

City of Three Angles
Explore maps in technology.

Every street in Three Angles was named after the shape of its corner. That made walking through town easy if you knew your shapes. Quad wanted to find Polygon Museum. He was new to the city, but he was pretty sure he could get around. Quad didn't bother taking a map.

Quad walked North along Line Street to the town square. He turned right at the square and walked East along Right Angle Road. Quad bought a cool dog collar at the Pet Palace. At the farthest end of Right Angle Road, Quad took a sharp right turn and headed South along Hypotenuse Street. He found two great bookstores along the way. Finally, Quad ended up right back where he'd started! He had not found the museum. He decided to try again on Friday.

The next morning, Quad looked for the Museum again. This time, he started on Hypotenuse Street. He headed North on Hypotenuse until the street came to a dead-end. Quad decided to turn right onto Acute Avenue. He was sure to find the museum at the end of Acute Avenue. Quad reached the corner and took a slight right onto Obtuse Place. He walked and walked until he saw the sign for his hotel at the end of Obtuse Place, just before Hypotenuse.

Good grief! The sun was setting. He'd have to find the museum on Saturday. This time, he'd save time and use a map.

On Saturday morning, Quad stood outside his hotel. The young tourist looked at his map. He touched the name of the Polygon Museum on the map. Slowly, Quad looked left. The museum was beside his hotel. Had Quad turned left out of the hotel, he would have seen the Museum right away. Next time, Quad would definitely use a map.

Traveling Triangle
Explore triangles in engineering.

Traveling Triangle
Explore triangles in engineering.

Triangle, Triangle,
where have you been?
I've soared in the sky
blown East and West by the wind.

Triangle, Triangle,
what have you seen?
I've watched the city
from the top of a Ferris wheel.

Triangle, Triangle,
what have you done?
I've held the wood steady
for builders in the hot summer sun.

Triangle, Triangle,
where have you gone?
I've biked along city streets
in the cool quiet dawn.

Triangle, Triangle,
what should we play?
Let's climb on the jungle gym
at the end of the day.

Fox's Forest
Explore environments in art.

Fox's Forest
Explore environments in art.

Fox trotted happily on the warm summer path. Sun beamed down through leaves to make moving patterns on the worn dirt track. Fox's head swung from side to side looking for yummy treats. He also scanned the woods for anything that might want to eat him!

Bushy tail wagging, Fox stopped to nose some flowers. Soon these green leaves would part to reveal little red berries. Delicious and sweet, Fox loved these berries in the straw. Fox caught another smell in the straw. It was the ripe odor of rotting log. Logs were good. They often held tasty crunchy bugs.

Fox lifted his little pointed face to the wind. Where was the smelly log? He faced West toward the setting sun. Sniff, sniff. No logs there. He faced East toward the far edge of the woods. No log smell rose from that direction. Fox turned North to sniff. There. The stinky log was that way!

Nose to the ground, Fox left the path and headed into the green woods. Shadows played across Fox's rust-colored body. Sniff, sniff. He followed the air molecules that blew in the breeze from that stinky log. The molecules were pulled in through Fox's nose and mouth. His brain made sense of the smells, and Fox turned a little to the West. Finally, behind a tall rock, Fox found a log so rotten it made his mouth water. Using tiny paws and claws, Fox pulled at the wet bark. Beetles and meal-worms tried to scatter. Fox caught them and crunched his way through an afternoon snack. Delicious!

Pointy little face covered in soil and bits of rotten tree, Fox turned back toward the path. South and a little East brought Fox right back to strawberry leaves beside the warm summer path. It wasn't quite so warm now that the Earth was turning away from the Sun. Shadows crept closer as Fox found his way back to the den. With a full belly, Fox curled up with his tail over his nose and fell asleep in the safety of his insulated, dry hole.

Tri, Tri Again
Explore triangles in mathematics.

Tri, Tri Again
Explore triangles in mathematics.

Poly really needed to fix her bird feeder's roof. The feeder hung from a branch outside her window. All kinds of feathered friends relied on the feeder for their winter meals. Sadly, there were a bunch of furry friends that also wanted to eat at the hanging seed source. Long fluffy tails knocked the feeder to the ground over and over again. Poly was afraid that the feeder would break in one of its squirrelly falls.

Poly had tried lots of different ways to trick the squirrels. She wanted to keep them away from the feeder, but the mammals wanted all the free seeds. Squirrels had nuts of their own, but the bird feeder was such an easy food source. Poly had designed a round tipsy cover that should have made the squirrels wobble and fall to the ground. They learned how to climb around it in less than a week. Poly had turned a mesh scarf into a wrap that squirrels couldn't get their hands through. They'd just untied and dropped the rectangular scarf to the ground. Poly had even covered the bird feeder hanger with drops of slippery oil. The greedy beasts just slid down it like a fireman's pole. They even seemed to have fun!

Poly had thought she was out of ideas. Round plates hadn't worked. Rectangular scarves hadn't worked. Spheres of oil hadn't worked. That afternoon, on her class trip to the city museum, Poly had seen a tall pointed roof. The roof covered a bell, and the guide told her class that the roof kept pigeons from landing. Bingo! If it kept pigeons away, the roof should keep squirrels away.

Poly cut out 3 triangles of wood. She covered them with waterproof material used for raincoats and connected all three triangles at their long edges. She formed a tall pointed object and attached it to the bird feeder. Her brand new triangular roof was nearly 3 feet tall and very steep. She hung the modified feeder and sat down to wait.

Birds flew in and out to get their seed meals. Finally, two fat little squirrels came to sit on the branch above the feeder. They tilted little heads to one side and the other side looking confused. Finally, one brave little friend leaped to the roof. She slid right past the seed and landed on the ground. That was one grumpy squirrel. Her rodent friends each tried the feeder, but all failed. Poly had learned her lesson.

If at first you don't succeed with circles and rectangles, tri, tri again.

NOTE: Tri indicates "three" and the spelling of the word that means to attempt something is "Try".

FLYING FIVES
Explore triangles in mathematics.

Once upon a time there was Rhombi. Rhombi loved shapes and found them everywhere. Her friends had helped her design a wheeled cart to carry things. She had a handy tool for checking a square's shape. She even had a plan for building strong bridges from triangles.

Rhombi was sitting on a very strong bench outside her school making the list for a school camping trip. Her team was exploring a cave soon and needed to drop supplies in the cave. They only had enough rope for themselves, and the supplies were really heavy.

Rhombi waved to friends. She saw a yellow school crossing sign by the road. The sign had 5 sides. She opened a search engine on her phone and typed in, "school crossing shape." The shape was called a regular pentagon. Rhombi typed "five-sided objects." The search engine returned a list of things made of regular pentagons. Because Rhombi always checked Internet facts, she went to look for pentagons on her own.

Rhombi sliced a star fruit, looked at a morning glory flower, and examined the surface of a soccer ball. Sudden motion in the sky caught her attention. A team of skydivers landed in the field. Their parachutes used fabric to slow fast falls to Earth from great heights.

This gave Rhombi an idea. Her cave supply problem might be solved with parachutes. Could she make one with a pentagon? Rhombi looked out from the pages of her world and asked, "Designers, would you help me create a pentagonal parachute?"

We're Hunting for a Pentagon
Explore pentagons in science.

We're Hunting for a Pentagon
Explore pentagons in science.

We're going on a 5 hunt. (We're going on a 5 hunt.)
We're looking for a pentagon. (We're looking for a pentagon.)
We're gonna catch a big one. (We're gonna catch a big one.)
I'm not afraid. (I'm not afraid.)
Are you? (No way!)

Here comes the crossing sign.
Better wait for the crossing guard.
Walk briskly and pump your arms.

REFRAIN

We're passing a tall house. It sure is wide and has 5 edges. Let's walk around it.

REFRAIN

We're walking through the soccer field. Uh-oh. Here comes the soccer ball. We better duck under it…

REFRAIN

We're running through the garden now. It sure is grassy. I see the morning glories and okra. Jump over the rows. Oooohhh. We're all muddy. Shake it off….

REFRAIN

I see a big building. It has 5 sides. It's The Pentagon.
Oh no, what time is it?
Recess starts in 5 minutes. We're going to miss it.
Better get home fast!
Oh no, mud! Shake it off.
Jump the okra and morning glories.
Yikes, duck the soccer ball.
Walk quickly around the tall house.

STOP! The crossing guard says to wait. Okay, we can cross now.
Walk briskly and pump those arms. Hurry! Hurry!
Ahhhhh, we made it!

Tool Time
Explore tools in technology.

Tool Time
Explore tools in technology.

Hilda and Poly had a job to do. Grandma wanted the garage organized. She liked everything in its place. The two friends had taken out the trash. They had bagged glass, cardboard, newspaper, and cans for recycling. They had even swept and dusted.

The tool bench was still a big mess. Tools were scattered on the table. Tools were dumped in boxes under the bench. Tools were hung in all the wrong pegs on the wall.

Hilda and Poly started by sorting the tools used to move items. They hung hammers and mallets on the peg board over jars of nails in lots of sizes. Star-shaped phillips head and flathead screwdrivers used to tighten or loosen things went into bins by size. #1, #2, and #3 were matched to their same sized screws.

Rhombi put grabbing, twisting, and pulling tools on the pegboard. Pliers were hung from smallest to largest. Wrenches were hung over jars of nuts and bolts.

Rhombi and Quad did not touch the cutting tools. Saws, hacksaws, pincers, and wire cutters were left for Grandma to handle later.

Carpenter's levels, rulers, angles, squares, and other measuring tools were hung neatly to the right of the pegboard wall behind the bench. A caliper used to measure very tiny items was placed in its box. Some tools were delicate.

When Hilda and Poly finished the job, they proudly called Grandma to see her shiny garage. The excited old woman clapped her hands and hugged both girls. When Grandma offered them money for doing the job, Hilda refused to take the money.

"No, Grandma," said Hilda. "I'm happy to help!"

Creative Design
Explore pentagons in engineering.

How Many Pentagons Can You Count In This Set Of Pictures?

Creative Design
Explore pentagons in engineering.

Pentagons have 5 edges, 5 edges, 5 edges.
Pentagons have 5 edges,
regular or not.

Regular pentagons have 5 edges
all the same, all the same, all the same.
Regular pentagons have 5 edges
all the same length.

Irregular pentagons have 5 edges
not the same, not the same, not the same.
Irregular pentagons have 5 edges
NOT the same size.

ALL pentagons have 5 angles, 5 angles, 5 angles.
All pentagons have 5 angles
regular or not.

The angles of all pentagons
add up, add up, add up.
The angles of all pentagons
add up to 540 degrees.

5 Edges
5 Angles
500 + 40 Degrees
PENTAGONS are fun!

Design Time
Explore polygons in art.

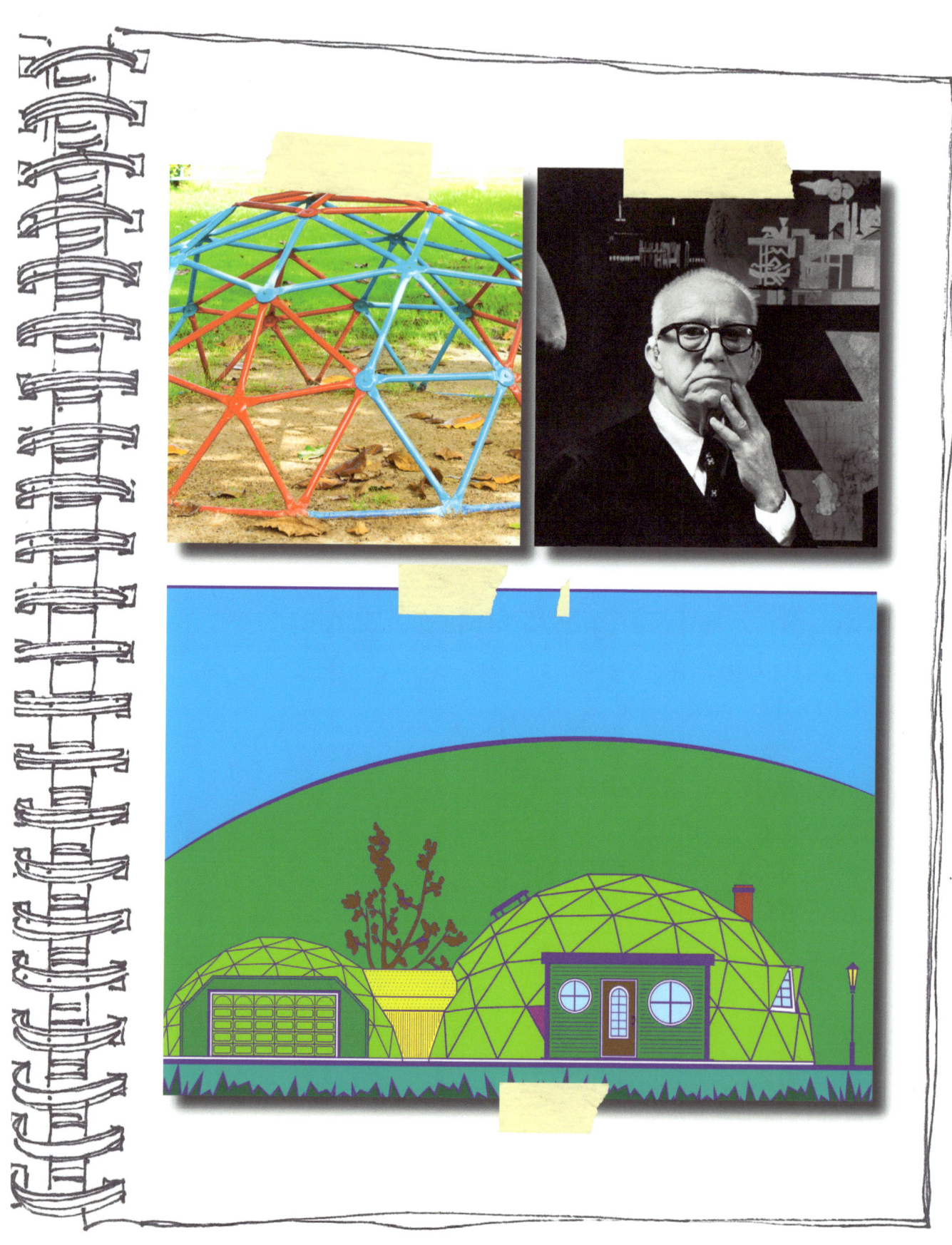

Design Time
Explore polygons in art.

Young man, you amaze me. *--Albert Einstein, physicist (1938)*

Buckminster Fuller said, "Whenever I draw a circle, I immediately want to step out of it." Buckminster Fuller was an architect and an engineer born in 1895. He was also a designer and an inventor. Fuller wrote 28 books on subjects in math and science. He had 47 special degrees from different universities.

Fuller was someone that cared about making the world better. He was also a humanitarian, someone that wants to improve life for other people. Each year the Buckminster Fuller Institute offers a $100,000 prize to "scientists, students, designers, architects, activists, entrepreneurs, artists, and planners from all over the world using innovative solutions to solve some of humanity's most pressing problems." From a floating doctor's office to a new plan for food production, Fuller's prize helps people with great ideas afford to put their plans in action.

This inventor was interested in using the planet's resources more efficiently. He invented a geodesic dome made of polygon shapes. The dome uses fewer materials and can be made bigger or smaller to meet the needs of many different types of uses. Gardens, homes, and museums around the world use the geodesic dome shape.

When Buckminster Fuller first imagined the special dome, he thought of it as humans working like a crew to survive on our planet, "spaceship earth." Epcot Center in Florida created a domed building inspired by Fuller's idea.

Domes like these are made of triangles. As more triangles are added, the shape gets stronger. No other 3-dimensional shape can be so big without extra supports inside the building. The larger it is, the stronger the dome becomes. These domes are so simple that an entire house can be built in a single day. They are so strong that geodesic domes can even survive hurricane force winds!

Can you see the triangles in the playground dome?
Can you see diamonds in the playground dome?
Can you see hexagons in the playground dome?

The Pentagon, USA

HOLDS A LOT OF CARS

 = 10,000 vehicles

Aerial Image by David B. Gleason from Chicago, IL (The Pentagon) via Wikimedia Commons

COUNT on the PENTAGON

1. formal dining room
2. cafeterias
3. days to design the building
4. thousand 2 hundred clocks
5. edges and vertices in a pentagon
6. snack bars
7. minute walk corner to corner
8. decades since building opened
9. hundred 21 feet per outer edge
10. thousand visitors a year

5 Stories Tall +2 below ground

5 Armed Services housed on site

5 Acre Central Plaza "Ground Zero"

Executive Order 8802 let The Pentagon open all its 284 bathrooms to all employees, black or white. For a time, The Pentagon was the only non-segregated building in Virginia.

NUMBER OF MONTHS FROM GROUND BREAKING TO OPENING CEREMONY											
	JAN	FEB	MAR	APR	MAY	JUNE	JULY	AUG	SEPT	NOV	DEC
1941									▇	▇	▇
1942	▇	▇	▇	▇	▇	▇	▇	▇	▇	▇	▇
1943	▇										

Reference Material
January 15, 2013 9 Things You May Not Know About the Pentagon By Barbara Maranzani via History.com
Vogel, Steve. The Pentagon: A History. New York: Random House, 2007. ISBN 1-400-06303-5 (pp. 199-200, 216-217)
History of the Pentagon — U.S. DOD info, archived at Archive.org.
Pentagon Web Site — By the U.S. Department of Defense.
http://en.wikipedia.org/wiki/The_Pentagon and Wikimedia Commons

Graphic Information
Explore info-graphics in mathematics.

Info-graphic is short for information graphic. An info graphic uses pictures and words to make lots of information and numbers easy to grasp. Look at this info graphic showing information about The Pentagon. You could read the paragraphs that follow, or you can use an info graphic to show the same information.

The Pentagon was designed in just 3 days and built far faster than most government buildings. It was also built to hold a lot of cars. Parking was planned for 10,000 cars. 16 months passed from the first day of construction until employees parked all those cars in the lots. The construction trucks, earth movers, and builders showed up for work on September 11, 1941. The Pentagon was completed in January of 1943. The building opened almost 8 decades ago.

The building serves a lot of needs. Very special visitors are invited to parties in the formal dining room. Most armed services workers eat in one of the 2 cafeterias. 5 different armed services use the building. 10,000 or more visitors use any of the 6 snack bars.

Finding a bathroom in The Pentagon is easy. When the building was planned in 1941, Virginia state law said that black people and white people had to use different bathrooms. The pentagon was built with 284 bathrooms. By order of the President, none of those bathrooms were ever really closed to any one group of people. Employees and visitors to The Pentagon all used any of the public bathrooms. For a time, The Pentagon was the only non-segregated building in the state of Virginia.

Getting from place to place in the huge building takes time. Each outer edge measures 921 feet. There are 5 stories full of offices and a 5 acre central area called "ground zero." The total square footage of the The Pentagon is more than 5 million square feet. Your school probably measures less than 50,000 square feet. Walking from one corner to another corner of The Pentagon takes nearly 7 minutes. How many seconds pass if you walk from one corner of your building to another corner of your building?

ON THE MOVE
Explore polygons in problem solving.

Once upon a time, there was Rhombi. Rhombi loved shapes, and she found them everywhere. Triangles showed up in benches and bridges. Squares were found in tiles on the floor. Circles helped people move from place to place. Even pentagons could be found in fruits and morning glories.

Rhombi was restless. School was out for the summer, and Rhombi wanted a change. She decided to pack her things for a road trip. Rhombi did not have a lot of money, so she would have to plan carefully.

Rhombi would need food, water, clothes, camping gear and a way to move all that stuff. She checked the map to decide on a destination. Where did she want to go?

Rhombi decided to head west toward the mountains and lakes. She drew her route on the map and chose a design for her vehicle. Builders had helped her, and she had a lot of plans from which to choose. She could make the vehicle even better this time.

Rhombi made her packing list for a 5 day trip. She included the food, water, clothing, blankets, and tent. She would need to stay warm, fed, hydrated, and dry.

Rhombi knew that she would need to cross the gap where a bridge had broken. That meant she'd need to carry materials for crossing in her vehicle. With all the designs given to her, she would need to choose the best bridge.

Rhombi also knew that she would need to drop her supplies from a great height before hiking down into a river gorge. The rolling vehicle would never make it down the path, and she couldn't carry all that stuff herself.

Which parachute design would she use? Finally, once she was at the river's edge, Rhombi would need to get her supplies across the water. Hmmm... Would she use a boat or a bridge or a fl ying contraption? How could she move all her things across the river?

Rhombi would need some help with this decision. She looked out beyond the pages of her world and asked, "Designers, will you use what you've learned about shapes to help me pack, travel, and cross the river with my supplies?"

SECOND GRADE MODULE 2: TABLE OF CONTENTS

RHOMBI'S HOUSE			58
SCIENCE	Wind on the Windows		60
TECHNOLOGY	Accidental Inventions		62
ENGINEERING	The Tiny House		64
ARTS	Cube World		66
MATHEMATICS	Do You Know the Quadrilaterals		68
UP ON THE ROOF			70
SCIENCE	Water Is as Water Does		72
TECHNOLOGY	The Case of the Disappearing Sand		74
ENGINEERING	Circle, Triangle, Square		76
ARTS	I Love Homes		78
MATHEMATICS	The Camel and the Pyramid		80
NEW HOME FOR PET			82
SCIENCE	Davis Finds a Purrl / Purrl Finds a Home		84
TECHNOLOGY	Rhombi Lived in a Zoo		86
ENGINEERING	It All Adds Up		88
ARTS	M.C. Escher's World		90
MATHEMATICS	How Big Is a Guinea Pig		92
PET'S CELEBRATION			94
SCIENCE	Honey, Honey		96
TECHNOLOGY	Waxing Colorful		98
ENGINEERING	Strength in Numbers		100
ARTS	Square in the Middle		102
MATHEMATICS	Coming up a Cloud		104
RHOMBI'S GALLERY			106
GLOSSARY			108

3D SHAPES

 6
 7
 8
 9

 EXTENSION

RHOMBI'S HOUSE
UNIT 6

RHOMBI'S HOUSE
Explore cubes.

Once upon a time, there was Rhombi. Rhombi loved shapes and found them everywhere. She especially loved the squares that made up each face of a cube.

Rhombi had just moved to a new town and needed to build a home. Rhombi chose a shape for her house that looked like blocks and sugar cubes.

Rhombi knew that she should build a sturdy house to withstand the fall winds, but the last warm days of summer called.

Rhombi hurried to finish her home so she could run and play in the sun.

That night, the weather changed. The air grew colder and blew wildly around the corners of Rhombi's home. Rhombi walked outside to see the storm for herself.

She jumped in surprise as one sharp gust of wind pushed at her house. Another gust of wind pulled at the house. The walls shuddered. The house shivered. So did Rhombi.

She stared at all the crispy leaves as they tumbled from the trees. Then, Rhombi stared in amazement as the walls of her house also tumbled to the ground.

Looking at the walls piled on the cold ground, Rhombi decided she needed some help. She looked out beyond the pages of her world and asked, "Designers, can you help me create a new and stronger house?"

Wind on the Windows
Explore weather in science.

Wind on the Windows
Explore weather in science.

My dog is scared. She is shaking and whining under the bed. I sit on the rug and rub her ears. It does not seem to help. Daisy is a big floppy German Shepherd, and she is not scared of anything. Well, she is scared of one thing. Daisy hates the sound that wind makes as it hits the corner of the house.

I try to tell her it is just the air moving. I speak calmly, and I stay very quiet and still. All these things help Daisy feel safe. I describe the sounds I hear. There are four tall pine trees outside the window. The wind through their needles sounds like a quartet of violins. Air pushes past the pine needles like a bow on violin strings. A willow tree by the creek trails its long branches into the water. Air lifts the branches and whistles through leaves like a dozen flutes singing.

Daisy is shaking a little less. I think she likes my story. Her head rests on my leg, and big brown eyes look up at me through tan lashes. Her tail thumps a bit. I keep talking.

Sound starts as vibration. I thump Daisy's name tag to make it jingle. Sound hits your ears like a pendulum swinging. Little tiny hairs in your ears start to vibrate. All those branches and leaves shaking outside are vibrating the air, see? Daisy's eyes roll to the window, but she does not lift her head. Faster vibration means high pitched sound. Listen.

We close our eyes and listen to the wind. All that moving air pounds the window panes, making each piece of glass rattle like a drum. Some of the wind slips through the cracks around the window, and we hear deep mellow sounds. "Those are drums and cellos in our symphony," I tell Daisy. She is not impressed.

Windy flutes, violins, cellos, and drums play. With the moon high above and the faded rug below us, Daisy and I drift off to sleep. Some wild and blustery version of Bach's Cello Suite No.1-Prelude flies through the night. Daisy calms down as we listen to the symphony of wind.

Accidental Inventions

1.0 POPSICLES

11 year old Frank Epperson wanted to make soda pop at home. In 1905, the popular drink could only be bought at stores and restaurants. Frank used his porch as a testing lab. He mixed powder and water to find the right taste and texture. Frank left the ingredients in a cup where he'd been working. Temperatures dropped overnight. Frank's drink froze with the stir stick stuck in its center.

Soda Pop + Icicle = Popsicle

2.0 WAFFLE CONES

The 1904 World's Fair is the birthplace of today's popular ice cream cone. Ice cream in dishes had been around for years. A dessert seller at the Fair was doing so much business that it quickly ran out of bowls and plates. A waffle maker at the next booth was not doing much business at all. The two business owners worked together to roll up the Persian waffles and fill them with ice cream.

Ice Cream + Waffles = Ice Cream Cone

3.0 CHOCOLATE CHIP COOKIES

The Toll House Inn's owner was named Ruth Wakefield. Ruth wanted to make some of her delicious chocolate cookies for people staying at the inn. Ruth realized that she was out of baker's chocolate. The inn keeper broke the regular sweet chocolate bar candy into little chips for the cookie dough. She expected the bits to melt into chocolate cookies. They did not. Instead, the chips stayed whole.

Dough + Chips = Chocolate Chip Cookies

4.0 STICKY NOTES

In 1968, Inventors Spencer Silver and Art Fry were researchers at 3M Laboratories. Spencer created a "low tack" sticky substance. It was unique in that it could be removed without hurting a surface. Unfortunately, no one had a use for the stuff. It went into storage. Many years later, Art Fry needed a way to keep papers in his choir's song book. He suggested a use for Spencer's sticky invention!

Low Tack Sticky + Paper = "Post-Its©"

5.0 SLINKY SPRING TOYS

Naval Engineer, Richard Jones, dropped one of the tension springs for a battleship project on the ground. It bounced from place to place all over the room. He spent about 2 years finding just the right materials and coil to turn his spring into a toy. Betty, his wife, named the new toy. "Slinky." Is a Swedish word that means sleek or sinuous. In 1945, the first batch of 400 toys sold out in 90 minutes.

Coil Spring + Hard Work = Slinky ©Toys

6.0 ?

What items do you use that might be combined to create a new invention?

Could you find a new use for something that you already own?

What would you invent to solve a problem or make the world a better place?

The Tiny House
Explore structures in engineering.

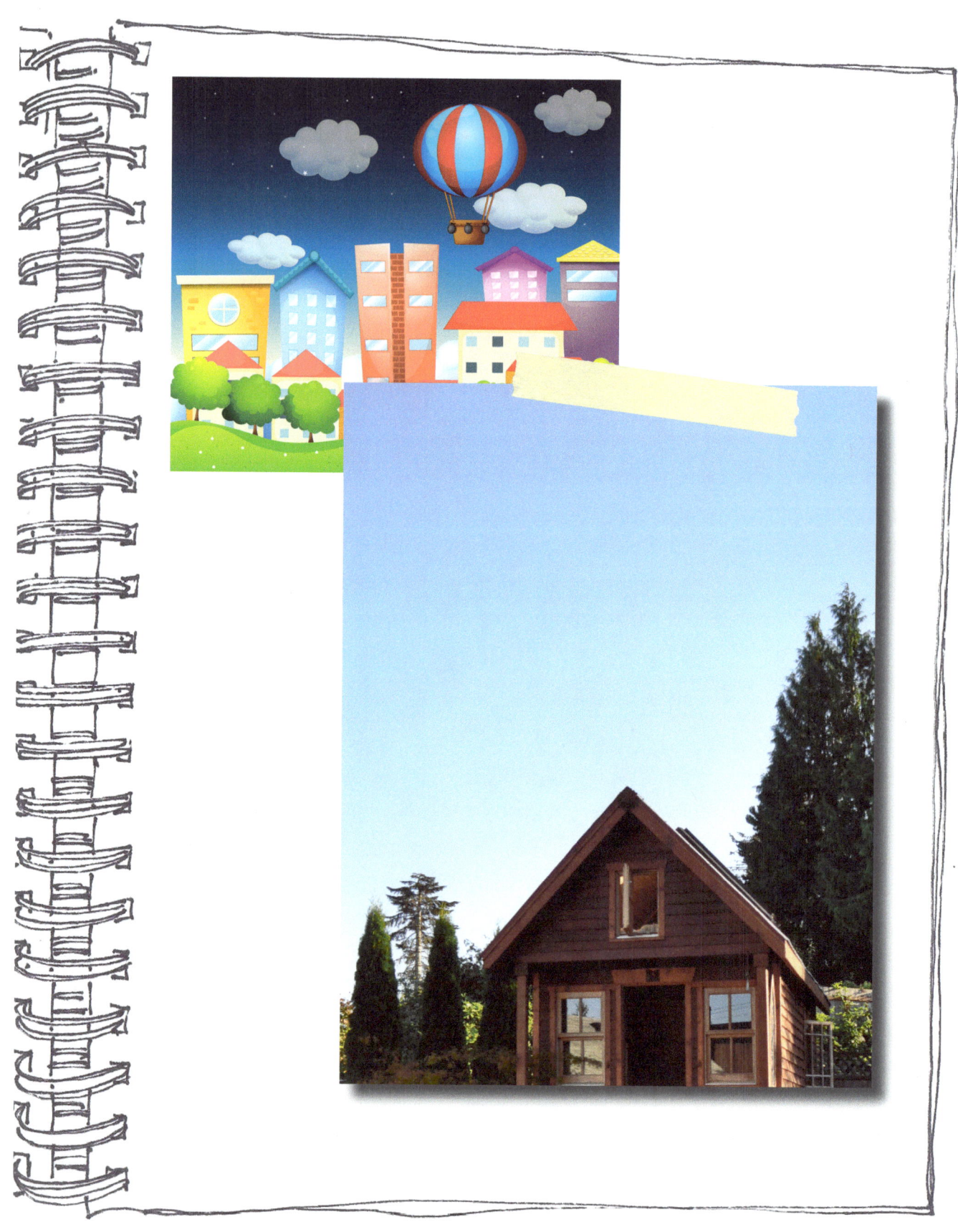

The Tiny House
Explore structures in engineering.

My Grandpa and I are building a tiny house. Grandpa says his tiny house will use less electricity and water than the house he and Grandma used to share. I am helping. Most of Grandpa's tiny house will be built far away and delivered to him in big flat boxes. We will put all the different pieces together to make a whole house.

There is a lot to do before the boxes get here. First we prepare the site. The land must be "graded." All the rocks and roots are taken away. The land is made as flat as possible. We drive a little road grader back and forth, then use a laser to level the ground. Next comes the foundation.

Grandpa forms and pour concrete cubes called pier pads. We have to wait a couple of days while the concrete sets. Two 4ft x 8ft beams rest side-by-side on the concrete. This holds the floor underneath the walls and house. I use a wrench to thread nuts onto bolts. Grandpa follows behind to tighten each one. He lays the drainage pipe and wiring through floor tunnels before finishing the sub-floor. Grandpa will use a composting toilet. All waste runs into a system of barrels under the tiny house to become fertilizer for his garden. The water will come from city plumbing, but grey water (waste water from sinks) will drain into the garden also.

Grandpa uses a generator to power tools for framing and roofing. Grandpa's walls were pre-made. Aunt Elizabeth comes over to prop up the walls and attach frames. We "plumb" the walls to be sure they stand up straight from the floor. I hold a "level" tool so that its bubbles are perfectly centered. We make sure windows and doors are level and square. We wrap it all in sealing tape to keep out water and mold.

Next, we'll add the roof. Grandpa uses a band saw to cut the rafters that rise from walls to roof. The rafters are spaced 4 feet apart. Insulation is added to walls and roof, then layers of roofing material top off the tiny house.

The house is done! Grandpa and I paint the tiny house and add shutters. We hang a sign that says, "Welcome to Grandpa's Tiny House in the Woods."

Cube World
Explore cubism in art.

Cube World
Explore cubism in art.

Artists began to look at the world as geometry in motion. Cubism began in France in 1907. Picasso and Braque started painting the world made up of cubes, spheres, cylinders, polygons, cones, and other geometric shapes. Cubists wanted to show all the sides of an object in the same picture.

The paintings looked a lot like the artist had cut the image and glued its pieces back together.

Cubist paintings show objects from more than one angle at once. Pablo Picasso and Georges Braque were first to add bright colors to cubist art. They believed that painters should not just repaint the world exactly as it exists. Instead, they wanted to show every part of the whole subject.

How does the world change when you re-imagine everything as two-dimensional and three-dimensional shapes?

Do You Know the Quadrilaterals
Explore quadrilaterals in mathematics.

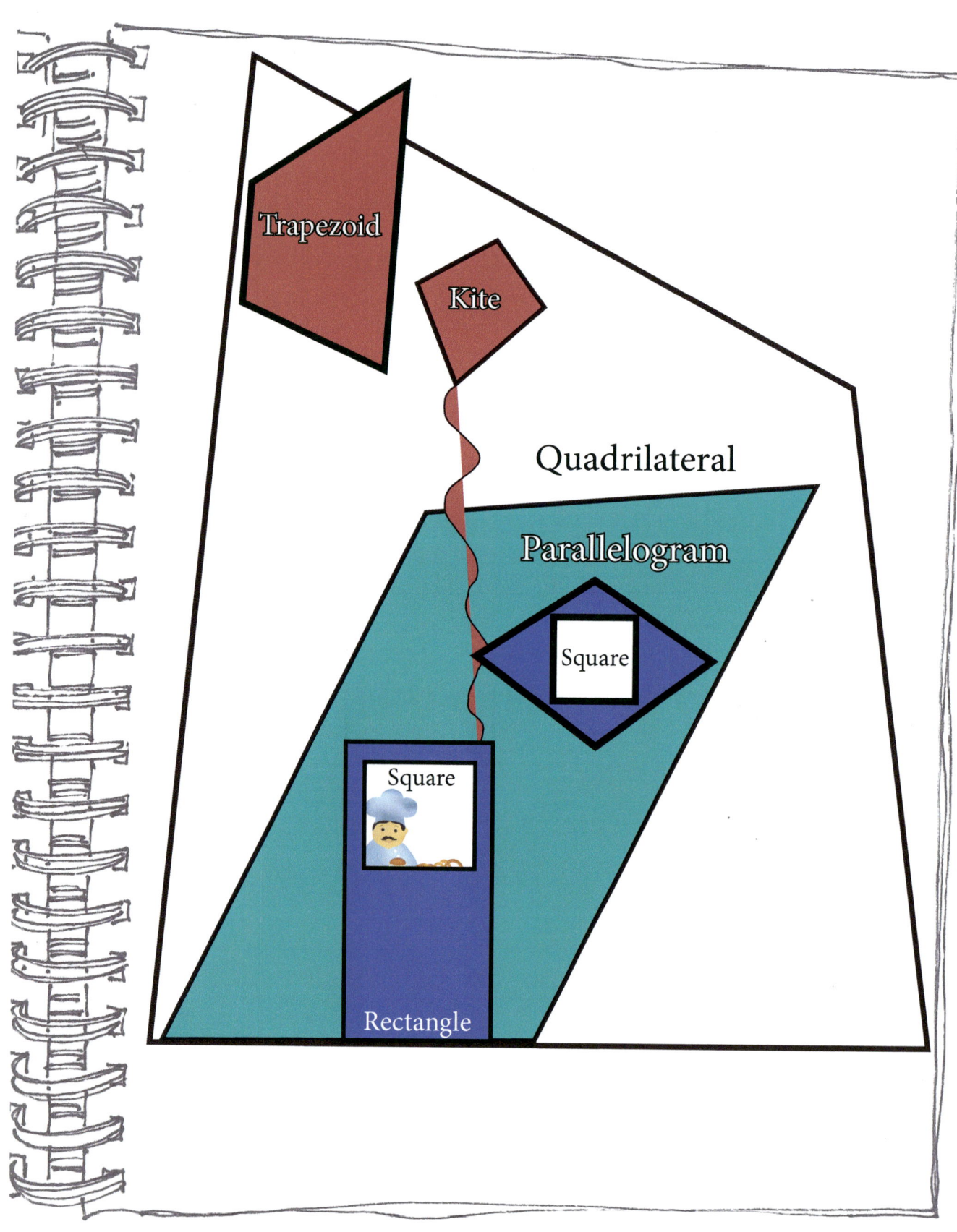

Do You Know the Quadrilaterals
Explore quadrilaterals in mathematics.

Oh do you know the quadrilateral, quadrilateral, quadrilateral?
Do you know the quadrilateral? A closed shape with 4 edges.
Oh, yes, I know the quadrilateral, quadrilateral, quadrilateral.
Oh, yes, I know the quadrilateral. 4 edges and 4 angles.

Now, do you know the trapezoid, the trapezoid, the trapezoid?
Do you know the trapezoid? It's a quadrilateral.
Oh, yes, I know the trapezoid, the trapezoid, the trapezoid.
Oh, yes I know the trapezoid. It has a pair of parallels. (a pair of LLs)

Now, do you know the parallelogram, parallelogram, parallelogram?
Do you know the parallelogram? It's a quadrilateral.
Oh, yes, I know the parallelogram, parallelogram, parallelogram.
Oh, yes, I know the parallelogram. It has TWO pairs of parallels.

Now, do you know the rectangle, the rectangle, the rectangle?
Do you know the rectangle. It's a quadrilateral.
Oh, yes, I know the rectangle, the rectangle, the rectangle.
Oh, yes, I know the rectangle… (deep breath)
It has two sets of parallel edges where opposite edges and opposite angles are the same.

Now, do you know how to find a square, find a square, find a square?
Do you know how to find a square. It's a quadrilateral.
Oh, yes, we know how to find a square, find a square, find a square.
Oh, yes we know how to find a square. Look for all edges and all angles the same.

Now we know our quadrilaterals, quadrilaterals, quadrilaterals.
Now we know our quadrilaterals.

Trapezoid, Parallelogram, Rectangle, Square!

UP ON THE ROOF
UNIT 7

UP ON THE ROOF
Explore pyramids.

Once upon a time there was Rhombi. She lived in a sturdy new house shaped like a cube. Her home was strong and safe. Autumn winds had not been able to push it down.

Rain was a problem, though. Winter rain and piles of snow were making pools of water on the flat rooftop.

Rhombi's friend climbed a ladder onto the rooftop to investigate. Drip. Drip. Drip. Freezing water and thawing ice had caused a hole in the roof!

The water had started to drip, drip, drip through the ceiling and into her home. If she didn't find a way to change the roof, her strong new home would be full of water.

Rhombi stood outside with her hot chocolate. She looked at her wonderful house and wondered what to do. She asked Radius and Pi for advice when they visited. The two thought a pyramid might work.

Rhombi decided to ask her friends for help. She looked out beyond the pages of her world and asked, "Designers, can you help me create a rain-proof roof for my house?"

Water Is As Water Does
Explore the states of water in science.

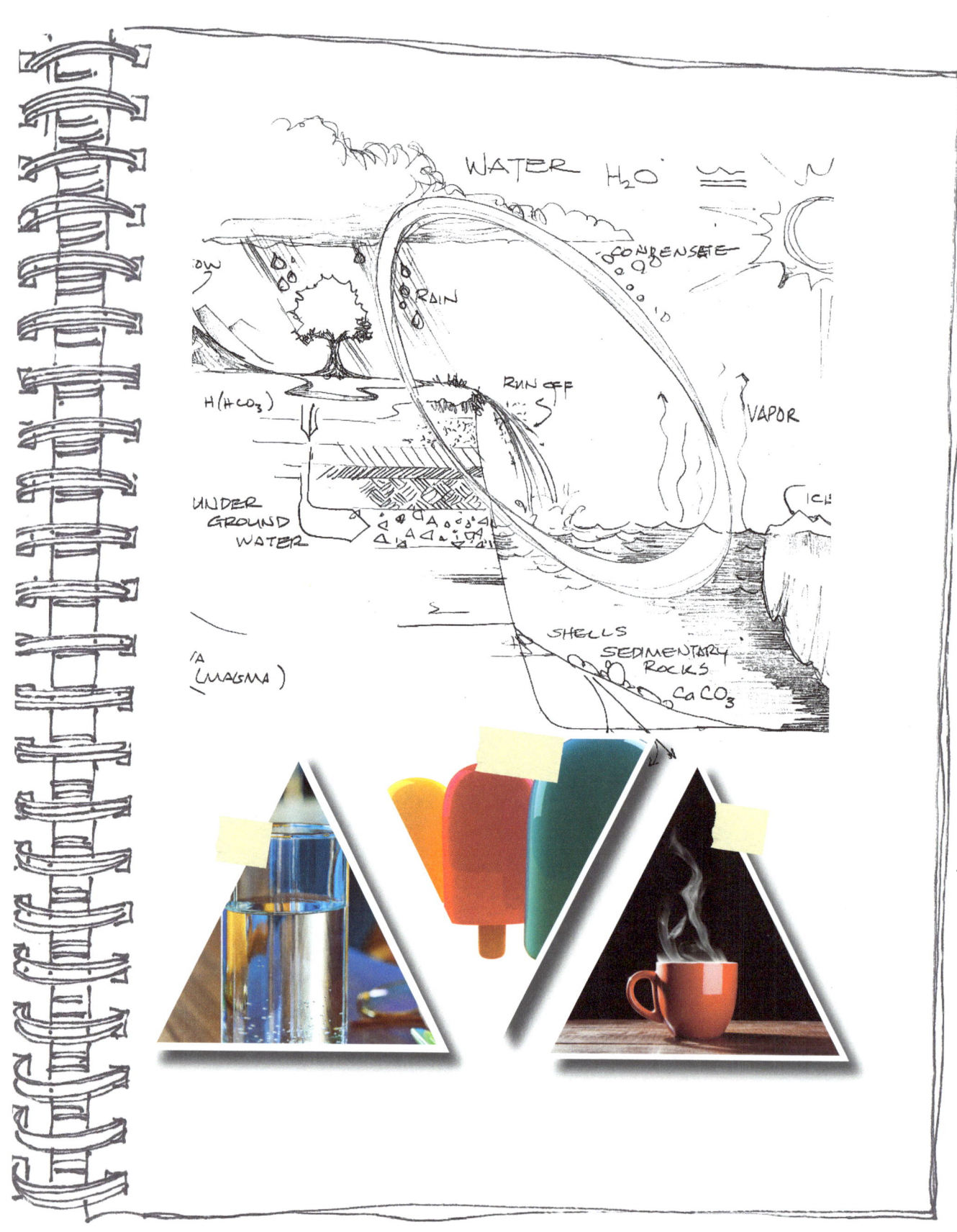

Water Is As Water Does
Explore the states of water in science.

LIQUID
Water that we drink - or spill - is called liquid.
Water changes names as it changes temperature.

ICE
Water that gets very cold turns solid.
Solid water is called ice.

VAPOR
When water gets warm enough, it seems to disappear but is still in the air. Water we cannot see in the air is called vapor.

WATER, WATER EVERYWHERE
How many of these phrases do you know?

Wet your whistle...
Like oil and water...
Mad as a wet hen...

CELSIUS
ice   0°
vapor   100°

FAHRENHEIT
ice   32°
vapor   212°

The Case of the Disappearing Sand
Explore the data collection in technology.

Read the charts with data collected by Rhombi and Pi.
What story does the data tell?
What happened to the sand sculpture?

DAY 1 THE PYRAMID BASE IS 3 FEET SQUARE 4 FEET HIGH SUNNY WIND 5 MILES PER HOUR
DAY 2 BASE IS ALMOST 4 FEET SQUARE 3 FEET HIGH SUNNY WIND 10 MILES PER HOUR
DAY 3 BASE IS 2 FEET ON THE LONGEST SIDE BUT NOT ALL SIDES ARE EVEN. 2 FEET HIGH ON ONE SIDE ALMOST 3 FEET HIGH ON THE OTHER SIDE (NO LONGER SYMETRICAL) RAIN LAST NIGHT WIND 25 MILES PER HOUR
DAY 4 THE BASE IS NOT SQUARE 1 FOOT HIGH ON ONE SIDE ONE SIDE HAS LOST A LOT OF SAND RAIN WIND GUSTING TO 4 MILES PER HOUR
DAY 5 MOSTLY FLAT AND NOT MUCH SAND LEFT

The Case of the Disappearing Sand
Explore data collection in technology.

The school bell rang. Kids tumbled out of the building. "What are you doing this weekend," Rhombi asked her friend Pi. Pi didn't have any plans. Neither did Rhombi. Maybe they would go for a swim or meet at the park?

Rhombi and Pi saw a bright orange flier stuck to the school bulletin board. The two friends looked at the sign. They looked at each other. They knew exactly what they'd be doing this weekend.

"Sand Sculpture Contest Saturday at 9AM"

Rhombi and Pi hauled buckets, spades, small shovels, and lots of cardboard boxes to use as sand molds. The team got name tags and were assigned a 5ft by 5ft area of sand. They sketched out a quick design. Rhombi used clay to mold a model of the sculpture. When both friends were happy with the design, they started to build. They dug. They piled. They formed and pressed. Finally, the sand started to look like a pyramid. Small camels milled around at the base, and sand tents held tiny people. About 4PM, Rhombi and Pi stepped back to look at their finished sculpture. It looked great!

Rhombi and Pi watched as judges made notes at each team's area. There were some amazing designs: seagulls flew over bridges; tsunamis crashed on beaches; castles held dragons, and a couple of robots held hands. The judges called all the contestants over to announce winners. Rhombi and Pi got bright green participation ribbons. They cheered as 1st, 2nd, and 3rd place winners were given medals. Then everyone headed home, tired and sandy but proud.

Rhombi and Pi passed the beach each day on the way to school. They noticed something interesting. The sculptures were changing shape. Each day, their pyramid design changed just a bit. The two decided to make a timeline that showed the changes every day for a week. Each day for a week, they made a fast sketch of the sculpture and jotted down notes to keep a good record. They also tracked rain with a measuring cup system and wind with an anemometer. The weather detectives could solve this case. What was making the sand disappear?

Circle, Triangle, Square
Explore rooftops in engineering.

Circle, Triangle, Square
Explore rooftops in engineering.

Circle, Triangle, Square
How many rooftops touch the air?

Round like the top of a castle turret.
Round like the top of a tower.
Round like the clock that's hand strikes each hour.

Slanted like the ridge of a Swiss chalet.
Angled like the top of a tepee.
Pointed like a spire in the city.

Flat like a hospital's helicopter pad.
Flat like the tar on a high rise.
Flat like the planetarium, mirrors pointed at the sky.

Circle, Triangle, Square
How many rooftops touched the air?

I Love Homes
Explore houses in art.

I Love Homes
Explore houses in art.

I love homes.
Tall homes… Small homes…
Round homes and square,
Homes in big buildings with hundreds of stairs.

I love homes:
Blue homes… Yellow homes…
Red homes and white,
Homes with black shutters to close out the night.

I love homes.
Prairie homes… Hobbit holes…
Towers and castles,
Yurts in the grassland and tents in the desert.

I love homes.
Stilt homes… Ranch homes…
Igloos and trailers,
Maybe I'll live on a boat like a sailor.

I love homes.
Cottage homes… Apartment homes…
Cabins or condos,
Tiny houses or duplexes lined up in neat rows.

I love everyone's home.

What kind of homes do you love?

The Camel and the Pyramid
Explore patterns in mathematics.

The Camel and the Pyramid
Explore patterns in mathematics.

There is a camel that lives in the far away sand of the Egyptian desert. His tan fur is stiff and rough. It has to keep out the tiny sand crystals as they crash through air during windstorms. He can close his nostrils and a third eyelid to keep out the terrible sharp sand. He has a hump that stores fat. The camel can live for days without food as his body uses the hump's fat to stay alive. Camels are tough.

The camel's caretaker is 14 years old. The young camel's guide wears a kafiyeh, a head wrap that does the same job as his camel's rough hide and closed nostrils. The scarf wound around the boy's head keeps sand and sun away from the boy's face. Wind blows lightly at 9 or 10 miles per hour through most days, but windstorms from the desert can be harsh. The wind may whip around the pyramids at 150 miles per hour. This kind of wind is dangerous to boys, camels, and pyramids.

The mahout knows that the pyramids will last a long time, though. As he guides tourists from all over world, he tells the history of these pyramids. He tells the families and backpacking teenagers that The Great Pyramid was made of horizontal rows of bricks. About 2,500,000 bricks were stacked like blocks. Each brick weighed around 2.5 tons. That means each brick weighed more than a car or a full-grown male giraffe. Tourists, happy with his stories, give him tips called baksheesh. The boy earns a lot of baksheesh to help care for his camel and help feed his family.

On this day in March, the winds are blowing. There are no tourists to hear his tales or ride his camel. The young mahout and the 800 pound camel head home to strong stables. They need to escape the sand and wind. As they plod away toward Cairo, the mahout looks over his shoulder at The Great Pyramid. It would not blow away in the storms, even though little bits of its faces were eroded by the constant wind. Thousands of years would pass before the wind and water slowly brought the pyramids to the ground.

The camel lowered his third eyelid, closed his nostrils, and headed home toward fresh water and good food and a dry place away from the itching, scouring winds.

ROOM FOR PET
UNIT 8

ROOM FOR PET
Explore habitats, polygons and structures.

Once upon a time there was Rhombi. She had a sturdy home shaped like a cube. A pyramid roof protected her house from the sun, wind, and rain.

One day in the Spring, Rhombi took a walk. She saw all kinds of pets with their people. There were puppies. There was a bunny. She even saw a guy walking with a lizard on his shoulder.

After seeing all those animals, Rhombi decided it was time to find a pet. She picked a scruffy little cat. Pet got a bath, a new collar, and a pillow on Rhombi's bed. Each night the tiny kitten curled up right under Rhombi's chin.

Things were great while Pet was little. As she grew, the cat needed more room to run and play. In the first week of April, Rhombi found Pet in the sink. "What are you doing, Pet," she laughed and scratched the kitty under her chin.

In the second week of April, Rhombi found Pet in the cabinet. Pet was a little sulky. She did not want to play.

Rhombi started to worry about Pet.

In the third week of April, Rhombi hopped to stay on her feet when Pet scooted across the room. Pet grumbled a little. In the fourth and last week of April, Rhombi tripped over Pet three times.

By the first week in May, Rhombi and Pet were grumpy about the tiny space. That week, Rhombi made a decision. For Pet's sake, she needed to give the kitty more space to play.

Rhombi realized that she needed a special place for her new pet. She looked out beyond the pages of her world and asked, "Designers, can you help me create a small house for Pet?"

Davis Find a Purrl
Explore perspective in science.

One rainy day, Meghan and Davis were splashing around in mud puddles. They saw a kitten curled up in a cold, sad little ball by the curb. It looked miserable and wet. Meghan tried to pick it up, but the kitten hissed and tried to scratch her arm. Davis pulled off his sweatshirt. They wrapped the angry kitten in Davis's warm sweatshirt. Thank goodness the kitten calmed and began to purr. It fell asleep.

Davis washed and brushed the little black ball of fur. He sat in front of the heater's vent with the kitten on his lap. It was so soft!

The veterinarian talked to Davis and Meghan about healthy food and habits. He made Davis promise to have the kitten registered and vaccinated. That would keep the kitten from getting diseases later in life. The veterinarian even helped pick the kitten food. Davis picked a soft food for nutrients and a crunchy food to keep the kitten's teeth healthy.

The kids named their new friend Purrl and got her a bright orange collar. Davis gave the kitten toys and a bed. He watched Purrl grow taller, and he noticed one day that his pet's eyes had turned green. Purrl got a patch of white fur grew on her chest and white "socks" on her paws. Purrl was growing quickly.

Purrl slept on the Davis's pillow every night and begged for breakfast scraps in the morning.

As Davis held his purring cat, he whispered, "They all lived happily ever after. Goodnight, Purrl."

Purrl Finds a Home
Explore perspective in science.

Once upon a time, there was kitten. She was lonely and cold. Rain fell on her fur and dripped in her sad blue eyes. She huddled at the curb trying to get out of the wind.

She saw two giants looking down at her. Cold wet dripped down from giant shiny bodies. Their giant hands reached for her, and she hissed. The kitten's fur stood up in tufts, and her tail stuck straight out behind her body. She hopped, trying to look big and scary. The world got dark and a little bit warmer. Still, the kitten shook and hissed and meowed. She struggled to get free of the dark dry thing, but it was warm and dry. The kitten calmed and began to purr. She was dry. She was warm. She slept.

When she woke, the giants petted her fur. She ate some yummy brown paste and crunchy cookie food. She drank clear, cool water. She curled up by a hot wind and fell asleep again.

The kitten woke to find a thing around her neck. She clawed at the thing. It made jingly noises. She chased it in a circle. The giants laughed. She hissed, but she didn't really mean it…

The kitten washed her fur every day. She grew as tall as the giant's knees. The giants got names. The one with long head fur that smelled like vanilla was Meghan. The one with short head fur that smelled like grass was Davis. Grassy David visited a lot.

Purrl slept most of the day waiting in the window for sun to touch her paws. That's when the big stinky thing spit Meghan out onto the sidewalk every afternoon. Then she and her Meghan played until bedtime.

Purrl purred as the two best friends settled in for the night.

Rhombi Lived in a Zoo
Explore habitats and data collection in technology.

Rhombi Lived in a Zoo
Explore habitats and data collection in technology.

There once was a Rhombi who lived in a zoo. She had 26 rescues and life was a hoot. She fed them all dinner with fruits, veggies, protein, dairy, and bread. Rhombi checked on each one before going to bed.

Anna the ape was tucked in her nest.
Bella the boa liked sun warmed rocks best.
Calvin the colt needed plenty of hay.
Dobbie the deer slept someplace new every day.
Ella the Eastern Gorilla slept under the stars.
Fred the ferret curled up in her arms.
Gina Giraffe slept standing up tall.
Hannah the hedgehog curled up very small.
Iggy the inchworm had a leafy bower.
Jasper the jaguar was just waking up at this hour.
Kara the kitten had a basket and blankie.
Her friend Latrice Lemur climbed a tree with her hankie.
Mable the mastiff took up a whole bed.
Nonny the newt liked her fishbowl instead.
Odetta Otter burrowed into a holt.
Polly the Pig gets a wallow or away she will bolt.
Qbert the quail must sleep alone,
But Rory the rabbit wants friends in his home.
Selena the sloth bunks down on a tree limb.
Tito Terrapin carries his bed around with him.
Una Umbrella Bird has a blanket in her 48 inch cage.
Vixen the fox prefers a crevice with sage.
Wally the wren chooses a nest up high.
XRay the tetra sleeps surrounded by bubbles all the time.
Yuri the yak flops in the dirt on his back.
Zen Zebra curls up in the hay.

There once was a Rhombi who lived in a zoo. She had 26 rescues and life was a hoot. Rhombi makes notes in her file each day. She updates the spreadsheet to show eating habits and play. Who is not sleeping? Who bumped its head? Which animals might want shade or sun instead? Rhombi finished her work, shut down her computer and crawled into bed. "I'll do it all again tomorrow," she happily said.

It All Adds Up
Explore 3D printing in engineering.

3D Printer

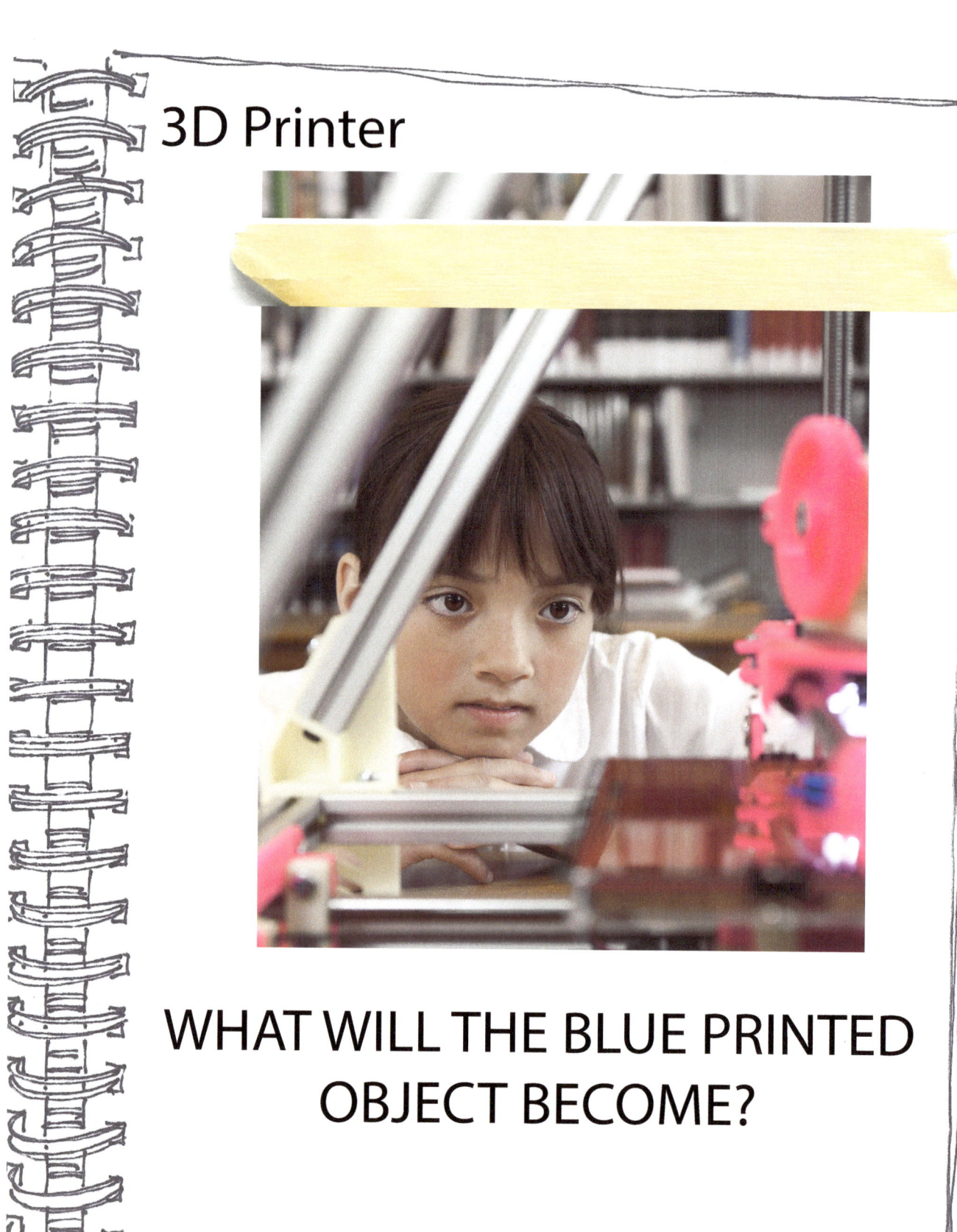

WHAT WILL THE BLUE PRINTED OBJECT BECOME?

It All Adds Up
Explore 3D printing in engineering.

Peg watched her dad add layer after layer to the party cake. Cake, icing, cake, icing, cake, icing and on and on until the cake was ready to serve. He slid the cake off its platform. He removed the little stilts that held the cake still as he iced the layers. Dad and Poly delivered the finished dessert to the museum for a party.

Dad dropped Peg off at school on the way back to the bakery. Peg ran to her classroom. Today, the new 3D printers were arriving. She was excited about designing for the cool little machines. When Peg arrived at her station in the classroom, she sat down at her wiggly table. She stuffed a piece of folded paper under one leg and tried not to let the table annoy her. After fixing the table, Peg found a note. It read, "Innovators look for new ways to solve problems. How will you improve the classroom using our new technology? Sketch. Model. Design. Create!"

Peg's teacher introduced the new printer technology, and explained that these machines were not like a 2D paper and ink printer. 3D Printers used Computer Aided Design (CAD) files to make computer models. 3D printing turns computer models into real physical things. They melt materials into thin layers on a surface, adding layer on layer until the full object is made. It sounded sort of like making a cake with Dad. He sketched and designed and layered. Leaning on her table, Peg's elbow slipped as the leg wobbled. "This table is so distracting," fumed Peg.

Suddenly, she had an idea. Peg got down on her stomach to take a closer look at the table leg. It was shorter than the other legs, and that made the whole table wobbly. Peg grabbed her logbook and started drawing. She remembered the little stilts her dad used to even out cakes for icing. She measured the length of the other three legs, and drew a sort of stilt for the table. Peg got modeling dough to make a 3D model of her idea. She asked for help in using the scanner and watched as the CAD file was created to turn her idea into a program the printer could read. Her scanned clay model was digitally sliced into thousands of layers in the CAD program. Each of those layers would add up to make her stilt.

Peg held her breath as the first layer of her "Peg Leg" design was melted onto the 3D printer's surface. Slowly, layer by layer, the object was made. At the very end of the day, Peg pulled her "Peg Leg" from the printer's shelf. She broke off the little filaments that held it to the shelf, and hurried to her table. Kneeling, Peg slipped the 3D printed plastic piece under the short table leg. No more wobbly tables! As she headed out to meet her Dad at the car, Peg wondered how she could make her "Peg Leg" even better. Maybe there was a way to change the height? Maybe she could design different shapes for square or round legs. The ideas were endless, and they just kept adding up…

M.C. Escher's World
Explore tessellations in art.

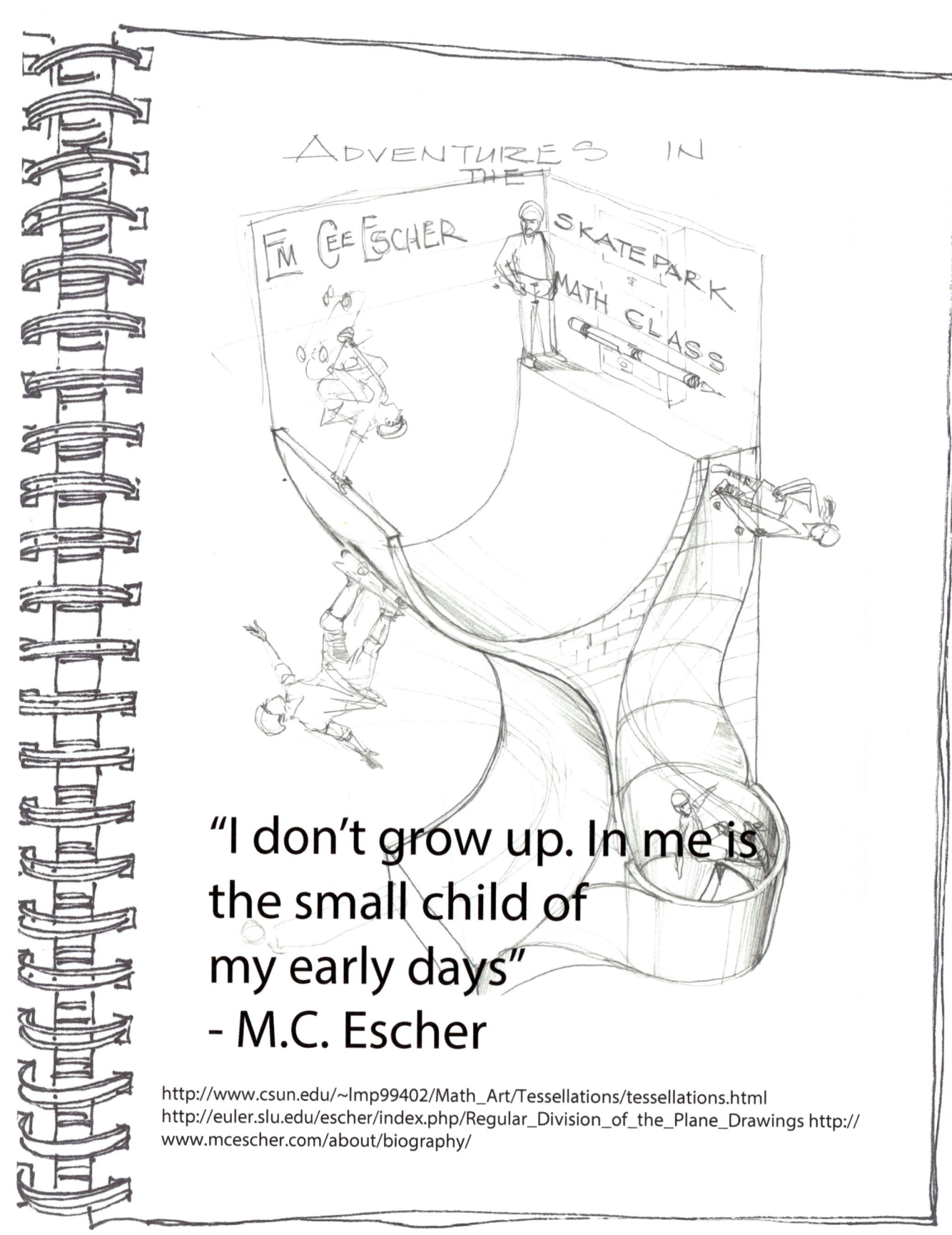

"I don't grow up. In me is the small child of my early days"
- M.C. Escher

http://www.csun.edu/~lmp99402/Math_Art/Tessellations/tessellations.html
http://euler.slu.edu/escher/index.php/Regular_Division_of_the_Plane_Drawings
http://www.mcescher.com/about/biography/

M.C. Escher's World
Explore tessellations in art.

Escher was a well-known graphic artist. He composed visual material for printing. Escher lived from 1898 to 1972. Like DaVinci and Michelangelo, M.C. Escher was a leftie. He used his left-hand to write and do most activities. Also like these great artists, Escher made a lot of art!

In his lifetime, the graphic artist made 448 lithographs, woodcuts and wood engravings and over 2,000 drawings and sketches. Escher also illustrated books, designed tapestries, created postage stamps and painted murals.

After visiting a Moorish castle called Alhambra, Escher became interested in the tiled floors and ceilings. He started drawing special patterns called tiling or tessellations. Escher filled five notebooks with 137 of these tessellation drawings.

Tessellations (or tiling) are created when a shape is repeated over and over again. Each shape fits right up against the next to fill an entire surface. The shapes do not overlap or leave space between them. He described ways to make a tessellation pattern.

You can make pattern pictures like Escher's tessellations using polygons. When you've practiced filling an entire page with geometric shapes, you will be ready to try animals, plants and trees.

ROTATION
turn around a center point

REFLECTION (FLIP)
flip over a line

TRANSLATION
move without rotation or reflection

How Big Is a Guinea Pig
Explore size in mathematics.

Labrador and Guinea Pig

How Big Is a Guinea Pig
Explore size in mathematics.

Food, Water, Air, Home.
Guinea pigs never live alone.

In groups of 5 or 10, wild guinea pigs use ½ inch claws to dig burrows in the ground for sleeping and keeping their tiny pups safe. Sometimes, the little community will use burrows left by other animals or bed down in rocky crevices. The guinea pigs work as a team, screaming to warn each other of predators or other danger.

Like rabbits, guinea pigs have incisor teeth that keep growing through their lives. Eating coarse grasses helps wear down the incisors. Premolars and molars crush and grind food. Also like rabbits, guinea pigs are actually rodents. They are not pigs at all!

Guinea pigs are mammals. Babies, called pups, drink mama sow's milk. Guinea pigs have fur, and they are born live and ready to wiggle. Pups grow quickly. From a few ounces to 2 or 3 pounds in less than a year, a full grown adult might be 10 to 14 inches long. That's about the length of lined paper or legal paper.

Guinea pigs live in their communities eating grass, sleeping through the day in safe burrows, and breathing the clear mountain air of South America from Columbia to Argentina.

Now that you know how guinea pig rodents live in the wild, how would you make a guinea pig happy in a classroom habitat?

PET'S CELEBRATION
UNIT 4

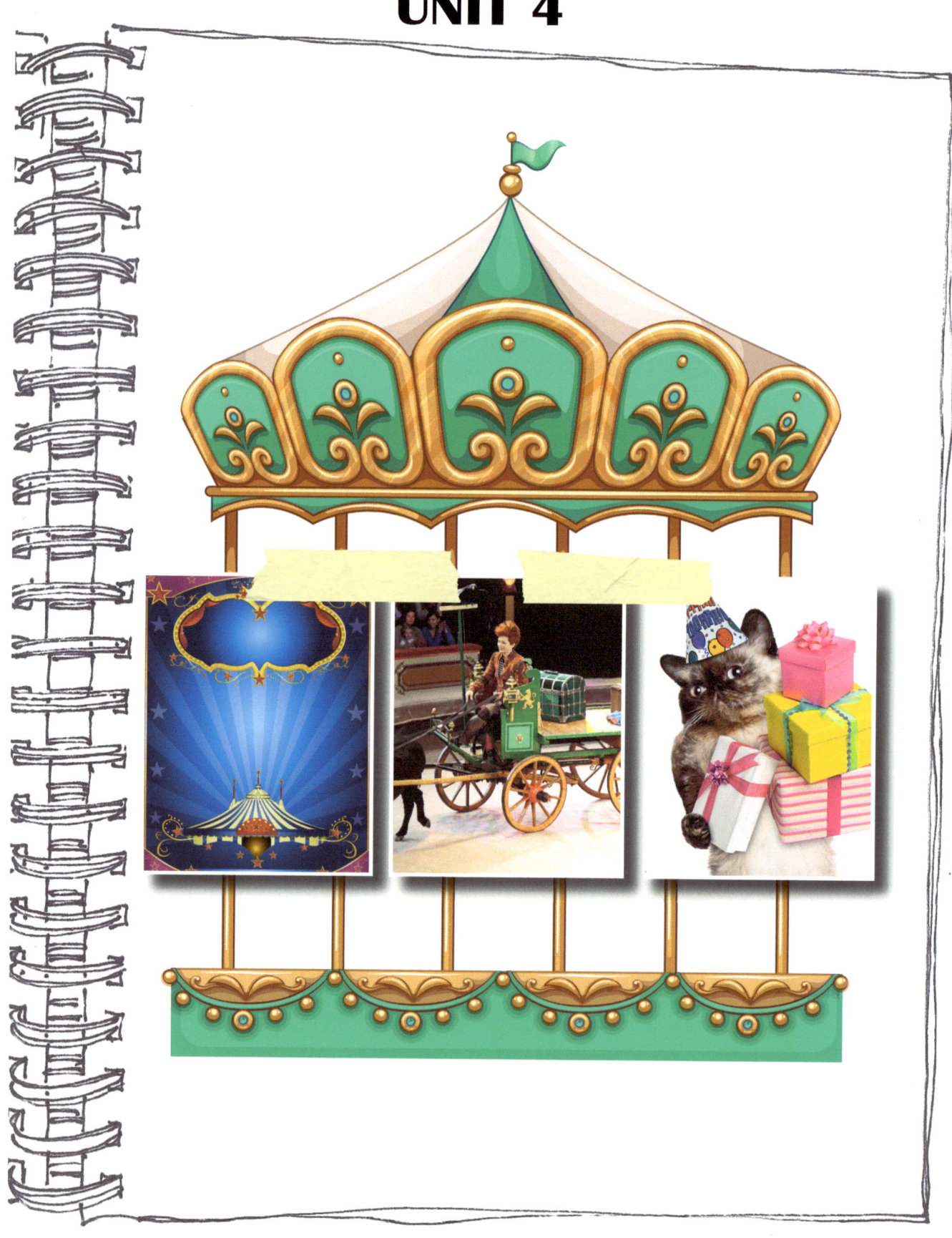

PET'S CELEBRATION
Explore habitats, polygons, and structures.

Once upon a time there was Rhombi. Rhombi had a great playhouse. Pet had a house, too. Both buildings would not be toppled by Autumn winds. The pyramid tops let winter rain and snow slide right off the roof.

Pet had lived with Rhombi for many months. The cat would have her 1st birthday on April 15th. After going to the circus in early April, Pet and Rhombi chose a "Big Top" party theme.

Rhombi was planning Pet's Summer birthday party and wanted to invite all of her barking, squawking, hissing, and mewing friends. Rhombi and Pet printed circus party invitations on primary colored paper.

Pet wanted 7 friends to attend her party. Rhombi invited a dog clown and ordered special cupcakes safe for pets! The third time Pet added a friend to the list, Rhombi realized that she didn't have enough room for a house full of 10 furry, scaly, feathery friends.

Each animal would need four square feet of space to be comfortable. She knew to ask for help. She looked beyond the pages of her world and asked, "Designers, will you help me design and build a covered space for Pet's birthday party?"

Honey, Honey
Explore bees in science.

Honey, Honey
Explore bees in science.

Honey buzzed about in a fancy pattern on the hive floor. She twisted and turned, bumping side-to-side and circling back to the start. She danced and danced until other bees began to head out of the hive. She lifted her head to watch the others through 5 eyes. On her thorax, four wings whirred into action. Honey's abdomen bumped the waxy honeycomb as she followed the other worker bees into the warm summer air.

Following her dance map, hundreds of honey bees flew to the garden where purple cone flowers, asters, blue forget-me-nots, yellow cornflowers, and bright fire-weed grew. Honey could not see red, but she could see the ultraviolet patterns in a bright red flower. She dropped onto a fire-weed and used her long proboscis to suck its nectar. Honey gathered pollen in a special pouch on her leg. As she landed on the next flower, some pollen from the fire-weed was bumped off her hairy legs. Honey carried almost 75 mg of nectar back to the hive in her honey stomach.

On the long flight back, Honey's antennae picked up the motion of a cat leaping toward her body. Honey darted back and forth to escape the animal. She landed in the hive to drop her pouch of nectar and pollen. When she was younger, Honey's older sisters had fed her with the honey they made. As Honey grew, she became the nursery worker and cared for larvae herself. Honey had helped with honey-making. She'd worked on making wax and building the honeycomb hive. Now, Honey's job was collecting the nectar that supplied carbohydrates and quick energy. She brought back pollen that gave the bees protein and fats they needed in their diets.

Honey paused at the opening of the hive. She checked for predators with antennae and her compound eyes. She used the simple eyes on top of her head to find the sun and pick a direction. Honey dropped from the hive to head out once more. She had many, many trips to make before the sunlight faded.

Honey bees were a little like nurses, a little like architects, and a little like builders. They created wax to build honeycomb in regular hexagons. The honey bees use a shape that fitst together as well as a puzzle. There are no holes or gaps between hexagons in the honeycomb. No wax-making honey or energy is wasted. That also made honey bees sort of like engineers learning the best way to do a job. The honeycomb "house" idea is working for humans, too. A company in the United Kingdom builds hexagonal houses. Each room of the house fits tightly together to save resources. These houses work for people the way a honeycomb works for bees. The honeycomb shapes take less energy to build, heat and cool. If you wanted to build a house, what shapes would you use?

Waxing Colorful
Explore wax in technology.

If you wanted to build a house, what shapes would you use?

Waxing Colorful
Explore wax in technology.

In the hive (or in a wild nest), there are three types of bees:

1. a single female queen bee
2. up to 2,000 male drone bees
3. some 20,000 to 40,000 female worker bees.

The worker bees raise larvae and collect the nectar that will become honey in the hive. When they leave the hive, they collect sugar-rich flower nectar and return.

Honey bees are a little like nurses, a little like architects, and a little like builders. They created wax to build honeycomb in regular hexagons.

The honey bees use a shape that fits together as well as a puzzle. There are no holes or gaps between hexagons in the honeycomb. No wax-making honey or energy is wasted. That also made honey bees sort of like engineers learning the best way to do a job.

The honeycomb "house" idea is working for humans, too. A company in the United Kingdom builds hexagonal houses. Each room of the house fits tightly together to save resources. These houses work for people the way a honeycomb works for bees. The honeycomb shapes take less energy to build, heat and cool.

People have many uses for the beeswax that bees will not use. They make a lot more than any hive needs. Bee farmers harvest the rest for things like candles. Wax from honeybees is also used to make crayons. Most crayons are made of paraffin, but some organic companies are making crayons that contain fewer chemicals. Beeswax is mixed with pigments to make a rainbow of colors. You might be using beeswax candles or crayons in your house!

Strenght in Numbers
Explore columns in engineering.

Strength In Numbers
Explore columns in engineering.

Ginny kicked up dust with her sneakers as she headed down the dirt road. She was just wondering if the day could get any hotter when a motion caught her eye. Something fluttered in the ditch to her left. She walked closer to get a look. She saw that it was most of a newspaper. Ginny did not want to leave litter in the road. She rolled the paper into a tube and tucked it under her arm to throw it away later.

She was about halfway to her friend's place when the sun broke free of the clouds. It was hot and sunny, and the glare made her head ache. Ginny remembered the newspaper. She took a page and folded a paper hat. Her class had made them for a play. She never thought that skill would be useful. Whew! Her face was cooler in the hat's shade.

Ginny pretended she was a pirate in her buccaneer's hat. She found a stick and heaved it around like a sword. Ginny was doing a pretty good pirate's "arrrr" when she crossed the little bridge over the Spring Road stream. Yesterday's rain had raised the water level, and there was a great current. Ginny stopped to unfold two pages of the newspaper. She folded two small paper boats and set them gently down in the water on the North side of the bridge. She ran to the South side of the bridge to see which boat would arrive first.

She turned back to the road and finished her walk to Meghan's house. Just as she turned into her best friend's driveway, Ginny saw Meghan's father looking for something in his car. Ginny was curious.

"What are you looking for, Mr. Harris?" asked Ginny. He showed her a letter that needed an envelope. Ginny had a solution. She used the last page of her newspaper to fold a quick envelope for Mr. Harris. He closed the flap with tape, added a stamp and mailed his letter.

He asked, "Why are you carrying that old newspaper around?"

Pirate Ginny answered, "Because, it's not just a newspaper..."

Square in the Middle
Explore mosaic patterns in art.

Square in the Middle
Explore mosaic patterns in art.

Quad sat glumly on the bench. There was a box on the ground between his feet. When Rhombi sat down, she looked in the box.

"Quad, why are you hauling a box of broken pottery and dishes around the school?"

Quad looked into the box and put his head in his hands. His elbows rested on his knees, and he groaned. "That was my birthday present for Granddad. I dropped it on the way out of the art room, and I don't have enough money to buy another one."

Rhombi pulled the box closer and pushed some of the shards with her pencil. She closed her eyes and began to brainstorm. What could you do with a box and broken plates? Rhombi thought about broken things. Where had she seen broken plates in the last week?

"I have it," cried Rhombi loudly. She giggled and said, "Sorry, that was really loud, but I have an idea." Quad's friend told him about the table her parents had bought for the patio. It had a pattern glued to the top in the shape of the moon and stars. The artist had glued each piece down and filled the spaces with ??.

"You could do the same thing with this box," said Rhombi. "Draw a design and fill it in with the broken pieces."

Quad sat up and grinned. He liked the idea. Granddad loved cats. Quad could draw a cat on the box. He grabbed the box and headed toward the art room. Turning, he said to Rhombi, "You are a genius!" She smiled and waved him away. Quad drew a cat. He glued all the broken pieces to the box. The art teacher gave Quad some grout to fill in the cracks between pieces. Mr. Kandinsky said that the grout would be dry in 24 hours. Quad left his masterpiece in the art room window. He was happy with the gift he'd give Granddad, and he knew Granddad would love that he'd made the present all by himself.

Comin' Up a Cloud
Explore materials in mathematics.

A cotton boll weighs about 4 grams. A paper clip weighs about 1 gram, so a cotton boll is equal in weight to **about 4 paper clips.**

Comin' Up a Cloud
Explore materials in mathematics.

Ginny carefully wove through the rows of cotton. The Georgia weather was hot and humid. Ginny felt like her clothes were sticky, and the tiny gnats made mad little dashes toward her face and ears. She fanned them away. They didn't bite, but the tiny flies were annoying.

She'd kept an eye on the cotton since planting season in February. The flower buds had opened, and she'd watched as the petals fell to the dusty ground. With petals gone, the rest of the plant had ripened and grown. Each plant had a lot of cotton "bolls" with fluffy fibers.

This week, she and her family would hire local help to harvest the bolls. Then the piles of not-so-fluffy cotton would be "ginned" to get rid of all the seeds. They would send the cotton off to a mill. The mills will mix and clean the cotton. Picking machines will break the piles of cotton into pieces. Dirt has to be shaken out of the cotton fibers.

A carding machine that separates the fibers will comb through the piles. Combing makes sure all the fibers run in the same direction. Ginny's Great-Great Grandma used to do all this by hand, but the machines work with a lot more fiber in a much shorter amount of time. Great-Great Grandma spent weeks doing what a machine finishes in an hour. A long, smooth rope called a "sliver" forms. The "sliver" has to be pulled and twisted until it's thin. The thin "roving" is finally ready for a spinning frame. Before it becomes string, all that beautiful roving is twisted and wound onto bobbins.

Ginny wondered where the cotton boll she touched would go. Would it end up on a cargo ship bound for China? Would it land in a pretty gift store? Would she wear it next year as a t-shirt or a pair of jeans? Maybe it would wind her yo-yo or fly a kite. As Ginny bent down to tie her cotton shoelace, a cloud crossed the field. She glanced at the sky. It looked like it was "comin' up a cloud."

Read more : http://www.ehow.com/about_6360791_cotton-made-thread_.html

RHOMBI'S GALLERY EXTENSION

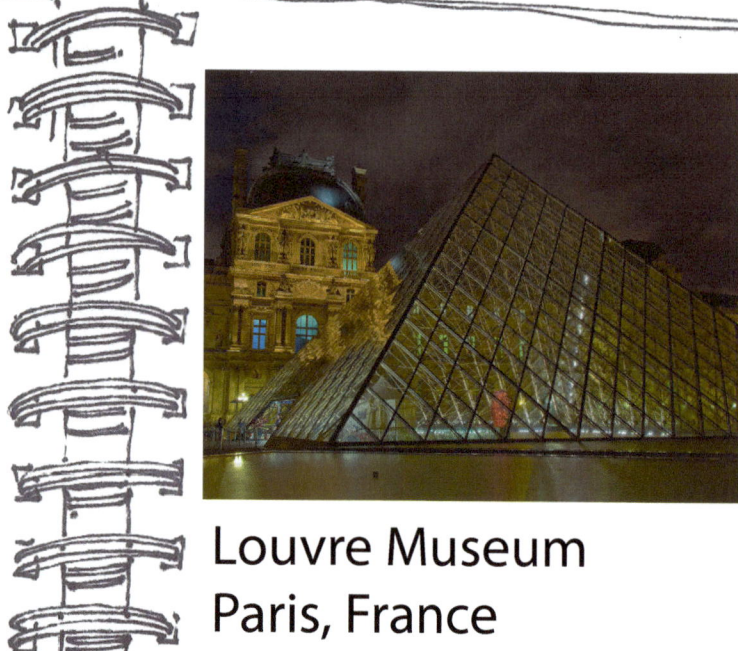

Louvre Museum
Paris, France

Guggenheim Museum
Bilbao, Spain

High Museum Exhibit
Atlanta, GA, USA

Brooklyn Children's
Museum
NYC, USA

RHOMBI'S GALLERY

Once upon a time, there was Rhombi. Rhombi had a wonderful friend named Pet and lived in a cool cube house with a strong pyramid roof. Pet even had her own room and a party place for friends!

Pet and Rhombi loved 2 dimensional and 3 dimensional shapes of all kinds. Their house was filled with interesting squares, triangles, hexagons, cubes, pyramids, and prisms. Rhombi's favorite was a frog in a trapezoid picture frame. Rhombi loved shapes so much that she was opening a geometric art gallery in town.

Rhombi and her friends were taking a leaf walk on their way to visit the site for Rhombi's new business. They found at least five kinds of leaves on the path and printed each one in Rhombi's sketchbook. Suddenly, Rhombi stopped and pointed to the empty and dark contractor's office.

Where were her general contractor, engineer, architect, and builder friends? They were all planning to meet for a brainstorm. Rhombi saw toolboxes on a table next to a set of blank blueprints. A note addressed to her was taped to the closed glass door.

"Rhombi, I've been called away to fix a building emergency. I can recommend a new architect and designer, or you can wait for me to return. I will be back in 3 weeks."

Rhombi looked at Pet and shook her head.

"The gallery fundraiser is set for 2 weeks from now. I need to have scale models for people to see, or I won't be able to raise enough money to build the gallery," said Rhombi. Pet swished her tail and meowed loudly.

"You're right, Pet. I should ask my designer friends for help.""

Rhombi looked out from the pages of her world and asked, "Designers, will you help me create the coolest art gallery you can imagine?"

SECOND GRADE
VOCABULARY

EARTH AND SPACE	
DATA	SCIENCE INFORMATION
conclusion	THE LAST PART
predict	TO SAY WHAT YOU THINK WILL HAPPEN
describe	SAYING OR DRAWING WHAT YOU SEE, HEAR, TOUCH, TASTE OR SMELL
observe	TO SEE, HEAR, TOUCH, TASTE OR SMELL
record	TO WRITE OR DRAW
identify	TO TELL WHAT SOMETHING IS
investigate	TO GATHER INFORMATION
evidence	INFORMATION OR FACTS THAT MAKE YOU BELIEVE SOMETHING IS TRUE
analyze	TO OBSERVE SOMETHING CAREFULLY IN ORDER TO UNDERSTAND IT
PHYSICAL PROPERTIES	
HAND LENSES	A HAND HELD MAGNIFYING GLASS
MASS	THE AMOUNT OF MATTER IN AN OBJECT
MATTER	ANYTHING THAT TAKES UP SPACE AND HAS MASS
NOTEBOOK	A BOOK OF BLANK PAGES FOR NOTES
OBJECT	THING
PATTERN	SHAPES, COLORS, OR LINES PUT TOGETHER IN A CERTAIN WAY
RELATIVE	COMPARING TO
TEXTURE	THE APPEARANCE AND FEEL OF A SURFACE
TOOLS	SOMETHING USEFUL
PHYSICAL PROPERTY	ATTRIBUTES THAT CAN BE OBSERVED AND MEASURED
WATER	
	THE CHANGE FROM A GAS TO A LIQUID
COOLING	TO MAKE LESS WARM
ENERGY	THE ABILITY TO CAUSE MOVEMENT OR CREATE CHANGE
EVAPORATION	THE CHANGE FROM A LIQUID TO A GAS
FREEZING	TO CHANGE FROM A LIQUID TO A SOLID BY LOSING HEAT
HEAT ENERGY	THE ENERGY OF MOVING PARTICLES
HEATING	TO MAKE WARM OR HOT
MELTING	TO CHANGE FROM A SOLID TO A LIQUID BY ADDING HEAT
TEMPERATURE	HOW HOT OR COLD SOMETHING IS
THERMOMETER	A TOOL USED TO MEASURE TEMPERATURE

SECOND GRADE
VOCABULARY

MOTION	
FORCE	PUSH OR PULL
MOTION	THE ACT OF CHANGING PLACE OR POSITION
PATTERN	A REPEATING EVENT
POSITION	WHERE AN OBJECT IS PLACED
SLIDING	TO MOVE FROM ONE PLACE TO ANOTHER WITH THE SAME PART OF AN OBJECT
TOUCHING	NOT BENT OR CURVED
ROLLING	TO MOVE FROM ONE PLACE TO ANOTHER BY TURNING OVER AND OVER
SPINNING	REPEATING MOTION THAT TURNS AROUND A POINT
FAST	TO MOVE QUICKLY
SLOW	TO MOVE AT A LOW SPEED
PROPERTIES OF MATTER	
MATTER	ANYTHING THAT HAS MASS AND TAKES UP SPACE
PHYSICAL PROPERTY	ATTRIBUTES THAT CAN BE OBSERVED AND MEASURED
SHAPE	THE OUTWARD FORM OF AN OBJECT
MASS	THE AMOUNT OF MATTER IN AN OBJECT
TEMPERATURE	A MEASURE OF THE AMOUNT OF HEAT
TEXTURE	HOW A SUBSTANCE FEELS
FLEXIBILITY	ABILITY TO BEND
SOLID	A STATE OF MATTER THAT HAS A DEFINITE SHAPE AND VOLUME
LIQUID	A STATE OF MATTER THAT HAS A DEFINITE VOLUME BUT TAKES THE SHAPE OF
CAREERS	
ASTRONOMER	SCIENTIST THAT STUDIES CELESTIAL PHENOMENA
CONTRACTOR	A GENERAL CONTRACTOR IS RESPONSIBLE FOR THE DAY-TO-DAY OPERATIONS OF A CONSTRUCTION SITE. CONTRACTORS MANAGE SALES PEOPLE, BUILDERS, PLUMBERS, ELECTRICIANS, AND ALL THE OTHER TRADES REQUIRED TO COMPLETE A PROJECT. THE CONTRACTOR ALSO COMMUNICATES INFORMATION TO ALL THE PEOPLE INVOLVED IN THE PROJECT.
DESIGNER	INDUSTRIAL DESIGNERS DEVELOP CONCEPTS AND SPECIFICATIONS THROUGH COLLECTION, ANALYSIS AND SYNTHESIS OF DATA GUIDED BY THE SPECIAL REQUIREMENTS OF THE CLIENT OR MANUFACTURER. DESIGNERS MAKE CLEAR AND CONCISE RECOMMENDATIONS THROUGH DRAWINGS, VERBAL DESCRIPTIONS, AND MATH MODELS.
ENGINEER	PROFESSIONA THAT APPLIES PRINCIPLES OF SCIENCE AND MATHEMATICS BY WHICH THE PROPERTIES OF MATTER AND THE SOURCES OF ENERGY IN NATURE ARE MADE USEFUL TO PEOPLE.

Log onto the website to access additional lessons, activities, word wall printables, printable nets, videos, downloadable files, links, and other resources.

www.ten80elementary.com

Please contact us for more information about curriculum, programs, festivals, and professional development.

info@ten80elementary.com